PLASTICS
The Layman's Guide

James Maxwell

Book 712
First published in 1999 by
IOM Communications Ltd
1 Carlton House Terrace
London SW1Y 5DB

© IOM Communications Ltd

IOM communications Ltd
is a wholly-owned subsidiary of
The Institute of Materials

All rights reserved

ISBN 1 86125 085 1

Typeset by Fakenham Photosetting Ltd,
Fakenham, UK

Printed and bound in the UK at
The University Press, Cambridge

CONTENTS

Preface	vii
Nomenclature	ix
1. THE STATUS OF PLASTICS An Overview of the Underpinning	1
2. BRIGHT NEW WORLD Innovations with Plastics	19
3. WHY CHANGE TO PLASTICS? The Pros and Cons of Replacement	33
4. SOMETHING FOR ALL OCCASIONS The Nature and Diversity of Polymers	49
5. TRANSFORMATION The Processing of Plastics	65
6. PUSHING POLYMERS TO THEIR LIMITS Composites	77
7. MAKING IT HAPPEN: PART ONE Identifying the Needs and Choosing the Material	93
8. MAKING IT HAPPEN: PART TWO Getting the Best from Each Material	109
9. FUTURE CONDITIONAL Challenges and Opportunities	123
Bibliography	139
Index	141

PREFACE

This book is for anyone interested in plastics. They may be professionally involved, they may be worried about the environment or fed-up with plastic packaging, or simply curious about a subject which was ignored when most of us were being educated. Those blessed with experience of the National Curriculum are unlikely to find material strictly related to any syllabus, but they may well enjoy an injection of the colour and flavour of real life.

There are no chemical formulae here, no engineering equations; just an occasional touch of polymer physics, where it helps to spread a little daylight. The emphasis is on the uses of plastics, on the multitude of applications in which plastics enrich (or complicate) our lives. There are thoughts about other ways things could be done; and there is some history, because this is so often the best key to understanding the present.

No book about plastics could avoid the big environmental issues of the day; questions about recycling, the use of resources and the like. The aim is also to give a proper hearing for the many unsung virtues of plastics, and the way in which they make possible things which were hitherto undreamed of. There is a search for balance, and for truth, insofar as either can survive undistorted in a simplified survey such as this.

NOMENCLATURE

If the polymer nomenclature in this book appears to be inconsistent and even chaotic, then that is the way things are. The failure of the Plastics Industry to provide simple, friendly names for its products amounts to a long-running Public Relations disaster. I can only blame my fellow chemists. Polymers are born out of very complicated chemistry; chemists have a deep-seated urge for precision, and throughout the early years the end-users had no idea what they were getting. So perhaps things could not have worked out any other way.

There are at least five ways of naming a polymer:

1. The systematic name. This is known only to the very young and the recently educated, and may be quite incomprehensible to the older chemist. Here polypropylene (PP) is poly(propene); polyethylene terephthalate (PET) becomes poly(oxy–1,2–ethanediyl oxycarbonyl–1,4– phenylene carbonyl).

2. The traditional name, like polypropylene and polyethylene terephthalate. These are acceptable to older chemists, and to engineers and others who have to work with them. However they are unfriendly to the public at large: in fact any word beginning with 'poly' is an undoubted 'turn-off'.

3. The familiar name, accepted by common use: nylon is the most familiar example. Alas, not many of these exist.

4. The abbreviation. PVC, ABS and PET are probably the best known. Other abbreviations are much less popular: PS (for polystyrene) is used, but can be confused with PS for polysulphone; PC (for polycarbonate) has too many other connotations, and PP (for polypropylene) is just undignified.

5. Trade names. These are not normally used, except where there is a direct commercial motive. But of course trade names when used to excess do pass into the language, as with Hoover and Kodak. With plastics it has happened with Celluloid, and is beginning to happen with Teflon (Du Pont PTFE) and Kevlar (Du Pont aramid).

The practice in this book is to use the appellation which is most acceptable to people who use the materials. Sometimes this will be an abbreviation like PVC, sometimes a 'familiar' like nylon or acrylic, and sometimes the full traditional name like polypropylene. So an apparent inconsistency is inevitable.

(When we stray away from Anglo-American, the 'familiar' names become less acceptable. In most other languages, the traditional names are preferred, hence polyamide (PA) for nylon and polyoxymethylene (POM) for acetal.)

1. THE STATUS OF PLASTICS
An Overview of the Underpinning

PERCEPTIONS AND MISUNDERSTANDINGS

Plastics are everywhere in the modern world. They contribute to every new technology; they touch everyone's life in every culture. Whole new industries have come into being, and undreamed-of effects have become commonplace. Many people (and not only the hundreds of thousands who work within the plastics industry) have a good appreciation of this vast contribution; others – including some opinion-formers within the media – do not. For them the impact of plastics goes uncherished, and often unnoticed, except for its negative side. As a result there is a serious imbalance in our awareness.

We can see part of the reason when we consider the things which have transformed the 'time and space' boundaries of life as our grandparents knew it. It is the ingenious end-products of engineering and electronics that have made their mark; the enabling materials scarcely signify.

This should not surprise us. The fact is that plastics did not initiate any of these seminal developments. Aircraft, automobiles, radio, telephones, vacuum cleaners, even the early computers; all these things took off without the aid of synthetic polymers. Natural polymers however were very much in evidence, with such essentials as hardwood for car body frames, doped canvas for aircraft skins, and rubber, gutta-percha and lacquered cotton for electrical insulation.

Synthetic polymers grew with these industries, gradually displacing the natural materials, as they surpassed them in performance, reliability and availability. The complexity and sheer scale of modern technology has confirmed the indispensibility of the synthetic polymers. But they did not initiate the process.

The main thrust of this book is concerned with polymers, formulated into plastics, for shaping by various processes into functional articles. However, the scope for polymers extends far beyond solid, three-dimensional items. Fibres, films, foams, coatings and adhesives are all heavily dependent on synthetic polymers. Fig. 1.1 gives an idea of the relative volumes of the polymer-based industries.

Frequently, a particular market niche is best met by a combination of traditional and synthetic materials. Milk containers in polyethylene-coated paper and the ubiquitous 65/35 polyester/cotton shirt fabric are good examples. Separately or in combination, polymers are established every-

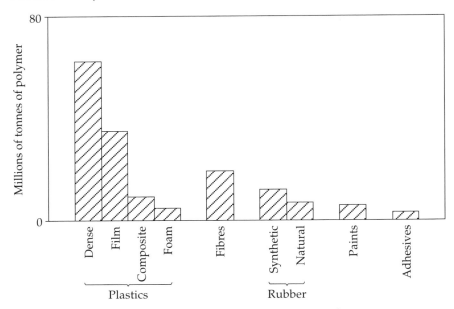

Fig. 1.1 The polymer industries compared

where. There is no going back: not without a catastrophic reduction in the scope and sophistication of life as we know it. Of course there are problems following in the wake of all these changes: indeed there is a formidable list of worries engaging us at the end of the century. We shall give them all an airing.

POLYMERS: THE BASICS

Our aim here is to achieve some understanding without recourse to the intricacies of polymer chemistry and physics. Polymers, (in fact chemistry in general) achieved a surprising amount of progress in former times without the benefit of a sound theoretical foundation. Nevertheless, real progress can only ride on the back of solid theory. The essential theory for polymers is the concept of long chain molecules. For any sort of understanding, the reader must be prepared to 'think molecules'. There is no dodging it, for it is this microstructure of long chains that makes polymers different from all other materials.

Astonishingly, the concept that polymers were made up of long chains or giant molecules did not emerge until the early 1920s, with Hermann Staudinger. Even then it was another decade before his ideas came to be generally accepted. Only with this understanding, and the new polymers

Plastics: The Layman's Guide 3

which followed on from it, was it possible for the huge superstructure of plastics-derived applications and effects to develop.

Different polymers are derived from particular chemical units (monomers): these (to use Staudinger's words) are the building blocks from which the giant molecules are formed. The particular characteristics of a polymer are decided partly by the chemistry of the monomers, and partly by the size and shape of the resulting molecules. Knowing whether the molecules in a polymer are straight or branched chains, or three dimensional crosslinked giants, can explain most of its properties. It often helps, especially in assessing the effect of heat or stress on a polymer, to visualise the molecules and the internal movement that may be possible between them.

Molecular structure explains the difference between the two main categories of polymer: thermosets and thermoplastics:

- Thermosetting plastics are formed by a chemical reaction involving pro-

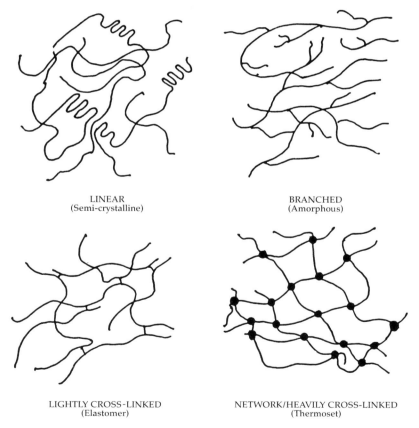

LINEAR
(Semi-crystalline)

BRANCHED
(Amorphous)

LIGHTLY CROSS-LINKED
(Elastomer)

NETWORK/HEAVILY CROSS-LINKED
(Thermoset)

Fig. 1.2 Diagram of molecular chains: Linear, Branched, Three-dimensional; Amorphous and Crystalline

gressive cross-linking in three dimensions. Shaping must be effected before this process is complete, because it irreversible.

- Thermoplastics comprise long chain molecules, essentially un-crosslinked. They can be softened and melted by heat, and then shaped before being allowed to cool and solidify. The cycle can be repeated many times without significant chemical change. Thermoplastics differ in behaviour according to whether they are crystalline or amorphous (see Fig. 1.2).

Table 1.1 lists the more familiar materials, embracing natural polymers as well as the synthetics. There is a third category, 'semi-synthetics', relating to natural products which were chemically modified for practical use. Most of these materials were developed during the busy early years of industrial chemistry, in the last century.

The other classifications of polymers, beyond this basic divide between thermoplastic and thermosetting, are discussed in chapter 4.

Table 1.1 Polymers – the basic sub-divisions

NATURAL	SEMI-SYNTHETIC	SYNTHETIC
--------------- THERMOPLASTIC ---------------		
horn	cellulose nitrate	polystyrene
bitumen	cellulose acetate	PVC
shellac	viscose rayon	polyethylene (LDPE & HDPE)
tortoiseshell		ABS & SAN
rubber		nylon
amber		polypropylene
gutta-percha		acrylic (PMMA)
		polycarbonate
		acetal
		saturated polyesters (PET & PBT)
--------------- THERMOSETTING ---------------		
lacquer	casein	phenol-formaldeyde
papier mache	vulcanised rubber	urea-formaldehyde
Bois Durci (albumen)		melamine-formaldehyde
cured bitumen		unsaturated polyester
cured shellac		epoxy
		alkyd

PLASTICS: THE LONG-FELT NEED

The comparitive 'invisibility' of plastics arises from the way they have appeared gradually, by displacement of natural polymers and semi-synthetics in established applications. The process of modifying materials to meet market needs was under way long before the advent of the petrochemical industry. Charles Mackintosh dissolved rubber in naphtha to produce the first waterprooof raincoats in 1823: Charles Goodyear's 'vulcanisation' of rubber with sulphur in 1839 led to a whole new class of workable hard materials in applications as varied as pipe stems and dentures. (Big events stimulated growth, as always: for example an order for 340 000 yards of rubberized cloth was recorded in 1853 for the British Army in the Crimea.)

New materials take over applications because the traditional products are either inferior or too expensive, or simply unavailable in sufficient quantities. The need for cheap, formable non-metallic materials had been building up throughout the middle of the nineteenth century, through the growth in population and spending power, and the demands of the new industries, particularly electrical and photographic. These were creating new applications, like insulation for telegraph cables (where gutta-percha provided the answer in the 1850s). The search for new materials was driven by the pressures on the supply of natural products like leather, horn, and shellac. One example of these pressures led to the famous 1860s cash prize highlighting the sufferings of the billiard ball trade, through the shortage of ivory. Not that the American addiction to pool tables was the main culprit: the pressure for a substitute for ivory knife handles was building up long before Parkesine and Xylonite (the British precursors of Celluloid).

Celluloid (originally the American trade name for cellulose nitrate) advanced very quickly once reliable production methods were achieved in the USA. Already in the 1880s it was in use for dolls and photographic film, and in the UK (as Xylonite) for knife handles and shirt collars and cuffs. Although there was little semblance of a polymer industry in 1914, it was rapidly harnessed, such as it was, to military needs. By 1918 celluloid had been pressed into service as accumulator cases, gas mask eyepieces, and accessories for machine guns, bomb fuses and the like. Cellulose acetate, being much less flammable, was preferred to Celluloid for photographic film and dope for aircraft wings. Casein, known to the ancient Egyptians as an adhesive, was used as a source of buttons and knitting needles.

Baekeland's 1907 patent for phenol-formaldehyde resin (PF), the first truly synthetic plastic, christened Bakelite, was not the culmination of a speculative project, but the result of a five year targeted search for a substitute for shellac. The only source of shellac is the secretions of particular Asiatic beetles; these were proving totally inadequate for the needs of the burgeoning electrical industry.

The pragmatic step-by-step growth of plastics continued in the first quarter of this century with improvements in semi-synthetics like casein and cellulose acetate. Arguably, the modern plastics industry does not date from the 1907 discovery of PF, but rather from the development in the 1920s of practicable methods for compression moulding its new filled modifications. This heralded the entry of plastics into mass production. Further landmark innovations were the first tentative injection moulding machine and its use with cellulose acetate, the foaming of rubber, and (just into the 1930s) the first extrusion applications for plastics.

The huge growth of the oil industry since 1945 has ensured security of supply, economy of scale, and (between the 1950s and 1970s), great proliferation in new materials. However, since the 1970s the rate of appearance of new polymers has fallen off dramatically: the field is now well ploughed.

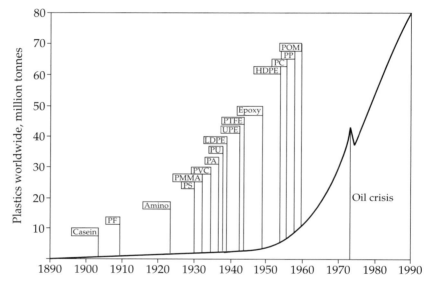

Fig. 1.3 The growth in the polymer market, with some of the milestones in new materials

The emphasis in the 1980s and 1990s has been on formulating existing materials to achieve focussed solutions to particular problems; on streamlining component design and tool design with computer technology; and on making processes more automated, more efficient and more cost effective. Gradually over the post-war period plastics have evolved from the 'long felt want' situation, in which they were welcome alternatives to increasingly scarce and expensive traditional materials, into something different. More and more nowadays they are being seen as new materials in their own right, performing in entirely new applications which could not have existed before.

WHERE THE POLYMERS GO

THE APPLICATION SPLIT

Subdivision of plastics according to application areas shows the preponderance of packaging, followed by the building industry. Other seemingly high profile application areas consume relatively modest amounts of plastics; in terms of material value, the balance is naturally rather different. Table 1.2 compares the size of the different end-use markets, which will be briefly surveyed in turn.

Table 1.2 Estimated end uses of plastics, by weight

Packaging	37%
Building and construction	23%
Electrical and electronic	10%
Transport	9%
Furniture	5%
Toys	3%
Housewares	3%
Agriculture	2%
Medical	2%
Sport	2%
Clothing	1%
Others	3%

PACKAGING

This is literally the most visible aspect of plastics. In Britain it accounts for nearly half of the total weight used, with a value around £2 billion per year. However it is not always the most favourably regarded manifestation. People are sceptical about the cost of packaging, and critical of the supposed waste of material. Plastics packaging is often seen as unnecessary: (after all, our parents managed without it!). These and other emotive issues are dealt with in chapter 9: here we can confine ourselves to looking at the benefits of this astonishing packaging revolution.

Food packaging is by far the most important sector. It is directly involved in the underpinning of modern life, in the very real sense that it guards and preserves the food supply. The function of packaging is to protect the contents from physical damage and contamination; from dampness (or, in some cases, drying out); from oxidation, and most important of all, from bacterial infection. The overriding statistic is that in the developed world, food wastage has been reduced to around 2%: elsewhere the absence of plastic packaging (and of adequate warehousing and distribution systems) can result in wastage as high as 50%.

The entire food chain benefits from LDPE (low density polyethylene), from black mulching film and baling film and translucent material for greenhouses through direct food packaging to supermarket carrier bags.

All the commodity plastics are used in packaging, in a variety of forms, frequently in association with more expensive polymers to create specific effects. Solid foods are sealed within appropriately impermeable films: predominantly LDPE film, but sometimes laminated to other materials such as nylon or EVOH (ethylene vinyl alcohol) to exclude oxygen from perishables such as meat.

Liquids, whether foodstuffs like milk, cooking oils or juices, or products like detergents, solvents or fuels are packed in bottles blown from a variety of plastics; in particular high and low density polyethylene, PVC (polyvinyl chloride) and PET (polyethylene terephthalate). The more temperature resistant polymers like PET and polypropylene are used for 'hot fill' products.

Another major sector is thin-walled packaging, for yoghurt, ice cream, desserts and the like. Foil in polypropylene or polystyrene is vacuum formed at high speed and the product sealed in with aluminium foil.

The driving forces here are shelf life and cost effectiveness, and the

Fig. 1.4 Film packaging; examples in metallised PET (courtesy Hoechst)

competitive pressure to provide a wide choice. The whole supermarket culture depends upon a guaranteed shelf life. Without modern packaging, food retailing would be far more restricted in choice, more seasonally dependent, and more expensive. The range of produce on the shelves has increased at least thirty-fold in the last forty years, in step with packaging developments. It would have been impossible for the industry to achieve its present sophistication in stock control and its astonishing variety of produce if its mainstream packaging were still limited to glass bottles, tinfoil and paper variants like kraft, 'grease-proof' and waxed card.

Plastics packaging reduces energy consumption in distribution, through weight saving, especially with blow moulded plastic bottles. PET is now the material of choice for all carbonated drinks bottles, and milk packaging, after many changes of direction, is now dominated by HDPE bottles designed in economical thin sections with built-in handles. Plastics also feature extensively in ancillary equipment such as aerosol valves, pump-action stoppers and child-proof and tamper-evident closures.

Foamed products play a big role in safe packaging and cushioning of fragile and expensive pieces of equipment. Expanded polystyrene (EPS) foam is spectacularly effective on account of its extreme lightness: it contains 98% air!

Fig. 1.5 PET bottle with pump dispensing unit in polypropylene (courtesy Lawson Mardon)

Plastic packaging has been a major player in the social revolution of the last fifty years. We have seen dramatic changes in work patterns, eating habits, shopping behaviour and leisure pursuits. Plastic packaging is woven into all these things quite intimately: how much as initiator, or facilitator, or by-product is an interesting question.

Building and Construction

Decorative high pressure laminates were one of the first high-profile applications of plastics in building. They appeared in post offices and Lyons Corner Houses in the early 1930s, and rode the oceans in the 'Queen Mary' and 'Queen Elizabeth' (the First!). They are still an important application. They are produced from combinations of paper with various thermosetting plastics, especially PF and melamine-formaldehyde, with familiar names like Formica and Warerite. Light weight 'honeycomb' structural panels (eg Holoplast) followed in the early post-war years. Sheet in acrylic (PMMA, polymethylmethacrylate) and polycarbonate (PC) is used in decorative panels and as protective panels over sports arenas, which also feature seating in nylon or polypropylene.

The most important polymer in the building industry is PVC. PVC in guttering, downpipes and all the other 'rainwater goods', mostly at the expense of cast iron: and PVC in window frames, at the expense of wood with its demands for repeated painting, aided by the cost benefits of double glazing. More than a third of European window frames are now in PVC.

PVC is also the market leader in that huge unsung market: pipes. Pipes are the invisible arteries of modern life: for fresh water, for drainage and sewage, for the gas supply, and now increasingly as conduits for electrical and fibre-optic cables for such things as power supply, television channels and motorway signalling.

The total usage for pipes in Western Europe is around 2 million tonnes annually. At least 70% of it is PVC, the other main materials being high and low density PE. Potable water pipes are usually made from PE, but there is still an alarming amount of lead piping in use. The much-publicised leakage from mains water systems in Britain arises from the age, fatigue and brittleness of the existing concrete, iron and clay pipes. The advantages of plastic pipes are very real. They last longer, they are much more resilient where there is ground movement, and the new twin-wall designs have very high impact strength. They are now being colour coded according to EC regulations (blue for potable water, yellow for gas, green for electrical conduit, etc). The principal factor however is the ease of installation. Their lightness makes manhandling much easier, fewer junctions are needed and joining is straightforward.

Plastics play a major role nowadays in civil engineering. Large areas of heavy duty LDPE sheet, assembled by hot gas welding, are used as water

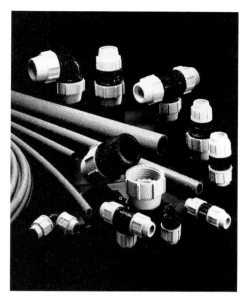

Fig. 1.6 One inch LDPE water pipe, with PVC fittings (courtesy BPF)

impermeable layers in major building and civil engineering projects like reservoir lining.

Beyond these high volume applications, plastics are to be found almost everywhere in the construction industry. Again, PVC predominates, especially in flooring and in exterior cladding, where GRP (glass reinforced polyester) is also widely used. The great proliferation of fixtures and fittings, both for the professionals and for the vast DIY industry, has been very largely built up on injection mouldings in polypropylene, nylon and acetal. Perhaps even more than the supermarkets, the DIY stores demonstrate the importance of plastics in our lives.

ELECTRICALS AND ELECTRONICS

Dependence on plastics extends right across the spectrum here, from the smallest item of electrical gadgetry in the home or the car to the heavy items of the supply industry itself. In its early years, the electrical industry was limited to rubber, gutta-percha and lacquered cotton for its insulating materials. The breakthrough into the array of modern formable and extrudable materials came progressively with PF, UF, PVC and LDPE. A powerful range of effects and functions is available to all of us now in the form of the huge range of appliances developed in the last fifty years. All of them depend upon plastics.

PF and UF (urea formaldehyde) still feature widely in basic domestic electrical hardware. Although its electrical properties are inferior, nylon is increasingly introduced into high impact or brightly coloured applications.

Material selection for electrical applications is conditioned by a number of mechanical and electrical parameters, over what can be an extensive temperature range. Cable covering, for instance, involves a wide range of thermoplastics: LDPE for heavy duty underwater cable, PVC for most telecommunications and domestic items, and more exotic (and expensive) low-flammability thermoplastics in aerospace and defence uses. Electric motors of all sizes use both thermoplastics and thermosets, usually reinforced, for end-frames and brush holders [Fig. 1.7].

Switches and connectors and electronic sensors, required in very large numbers, are injection moulded in engineering thermoplastics, the choice depending on temperature, chemical environment and dimensional tolerances. Notably, nylon is preferred for impact resistant snap-fit connectors, PBT (polybutylene terephthalate) for more dimensionally precise components, and a variety of high performance engineering polymers for high temperature applications needing low burning and low smoke emission characteristics.

Housings for all manner of appliances nowadays are made in plastics. Toughness is the prime feature of hand tools; polypropylene is the first choice for house and garden equipment, with reinforced nylon the likely alternative for sub-assemblies where form stability at high temperature is important. Housings for TV sets and business machines, computer monitors, keyboards and printers all require extra dimensional stability: amorphous plastics such as ABS (acrylonitrile butadiene styrene) and blends based on PPO (polyphenylene oxide) are the likely choice. For tool housings, plastics offer double

Fig. 1.7 Brush carrier for electric motor used at high operating temperature, in nylon 46 (courtesy DSM Polymers)

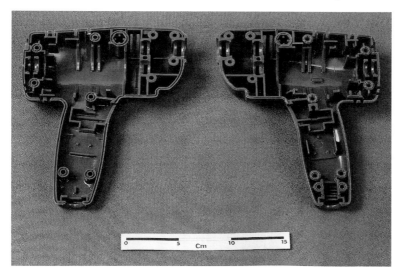

Fig. 1.8 The original Black & Decker drill housing in glass fibre reinforced nylon 66 (ICI, 1964)

insulation, a clear safety advantage over metal die-castings: the possible disadvantages of static build-up and electromagnetic interference can be overcome with special formulations.

TRANSPORT

Electrical components also feature in all forms of transport, right across the spectrum from the humble bicycle lamp to aircraft direction finding equipment. In all the more sophisticated forms of transport, electronics are exercising a more and more crucial role, giving more opportunities for high temperature engineering plastics.

Different parameters operate in the individual transport sectors. In the aircraft industry the dominant considerations are weight saving, because of its critical influence on payload, and safety, in its various manifestations of flammability, smoke evolution and impact resistance. Structural composites are already in use for wing sections and tailplanes, and even in complete airframes. Injection mouldings feature in numerous mechanical components, and sheet materials are extensively used for internal cladding: in all cases the flammability specifications limit the selection to relatively elite and expensive plastics.

Rail transport has for many years been using nylon mouldings for track insulation and GRP shapings for seat frames and exterior body sections. New designs and increased demands for weight saving and low smoke, low toxicity burning performance have led to the introduction of new materials like modified acrylic resins, both for seating and cladding.

In transport, the main focus for plastics is of course the automotive industry. A typical car contains around 1000 plastic components, out of a total of around 3000. The weight of plastics often exceeds 100 kg; typically 11% or 12% of the vehicle weight. This all adds up to a global market of over 4 million tonnes of plastics, valued at more than £5 billion in raw materials alone.

Significant weight reductions are now being achieved in vehicle weight through the introduction of plastics parts, offsetting weight increases due to additional functions and safety features. Although weight reduction is a real benefit in terms of vehicle performance, fuel conservation and exhaust emission, it has rarely provided the main stimulus for the use of plastics. The usual stimuli are the cost savings arising from reductions in the number of components and simplified assembly operations. Additionally, over the last thirty years the design potential of plastics has generated spectacular innovations. Prominent among them are the resilient front and rear ends (most frequently polypropylene backed with polyurethane foam), and the fuel tank in HDPE, less lethal and more space-saving than the traditional pressed steel tank.

In the car interior almost everything visible, except the glazing, is a synthetic polymer, either as solid polypropylene, PVC or ABS, or as nylon or polyester fabrics. Polyurethane features as seat upholstery and visibly as self-skinning systems for steering wheel and gear shift.

In the engine compartment, the cooling, climate control and fuel supply systems are heavily dependent on reinforced nylon and polypropylene, along with one-piece front-end structures. Developments like the intake manifold in glass reinforced nylon are representative of leading edge technology, in material development, processing methods and production engineering.

Fig. 1.9 Modern passenger compartment: Toyota Previa (courtesy GE Plastics)

Medical Applications

Plastics may not be consumed in huge quantities in medical applications, but they are certainly indispensable. Interestingly, they feature at both ends of the technology spectrum. Plastic prostheses are essential in hip and knee joints and in heart valves. High performance materials are included in such equipment as dialysis pumps and X-ray diagnostic machines. Nevertheless it could be argued that the day-to-day underpinning of hospital activity with cheaper polymers like PVC, LDPE and polystyrene, as disposable blood-bags, tubing and syringes and other nursing 'hardware', constitutes an even more vital contribution. There are specific benefits also from plastics packaging, in such items as blisterpacks for medicaments and childproof closures for bottles.

Dentistry has been a fertile field for plastics since the earliest days. The connection continues, with the development of a variety of new quick-curing thermoset resins for fillings and dentures.

Sport

In sport, people extend themselves. At the higher levels of competition, they push themselves to the limit. The same thing happens to their equipment. Hence we find nowadays that sporting goods become a test bed for materials. The demands made of polymers, in fabrics and in solid components, become more and more severe. It is not difficult to appreciate the reasons for the popularity of plastics in sport. High strength and rigidity coupled with lightness, high impact strength, low friction and lubricity, good resistance to

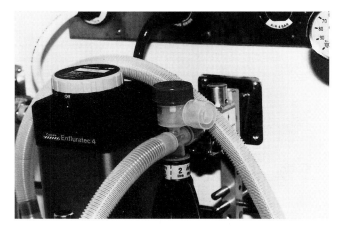

Fig. 1.10 Single use anaesthetic breathing system, with hose in PVC, body in polypropylene, and critical components in acetal copolymer (courtesy Hoechst UK)

16   Plastics: The Layman's Guide

Fig. 1.11 Sporting applications

tearing and abrasion, 'breathability' and water resistance: the list is formidable. Low cost foams are used in abundance; but, like medicine, sport is a market where high added values are tolerated. Hence carbon fibre reinforcement is used in thermosets for demanding designs of golf clubs, fishing rods and vaulting poles, and in nylon for injection moulding of top-of-the-range tennis racquets.

OTHER APPLICATIONS: HOUSE AND GARDEN, FURNITURE, TOYS, ETC.

The logic of lumping these very different categories together is that they share common requirements: toughness, resilience, colourability, abrasion resistance, low cost, and survival without maintenance.

Polypropylene is the workhorse that is transformed into those hundreds of taken-for-granted applications: handles for garden tools, containers and tools for the kitchen, crates and tote boxes and toys of all description. All are tough, scratch resistant and usually moulded in bright primary colours. The large, blow-mouldable or rotamoulded items, be they garden kneelers or wendy houses, are usually HDPE. Nylon is more discreet, less conspicuous (and indeed, more expensive) but serves as the lynch-pin of the flatpack furniture business: all those self-assembled units are held together with small nylon mouldings.

Plastics have attracted the attention of furniture designers from the early days. The new materials were seen to offer an escape from the traditional base-plus-back-plus-legs concept of kitchen and dining chairs. Some very innovative solutions have emerged over the years, but the one which has taken over the mass market is injection moulded polypropylene, in the form of a single shaping mounted on steel legs. Garden furniture is well and truly the preserve of polypropylene, usually in weather-resistant white or dark green compositions.

SUCCEEDING WITH PLASTICS

Plastics are employed for the benefits they impart: better styling, weight reduction, labour saving, cost reduction, or some entirely new combination of effects. There may be those who regard them as inferior substitutes, inflicted upon a reluctant public. But in fact the onward march of plastics has always been much more a process of 'market pull' than one of 'product push'. We buy plastics because we want what they can provide. Often it is a very basic need, not merely the froth on an over-affluent society. All too frequently when we see the results of some natural or man-made disaster on our TV screens, the misery is highlighted by a glimpse of mundane, everyday things, in brightly coloured plastics. Often they are seen in a life-sustaining role, as LDPE groundsheets, or polypropylene buckets, or HDPE bottles or PVC sandals.

Of course, as always, success brings its problems. Discarded plastics packaging is very visible, in our dustbins, on our streets and on our beaches; a litter problem which is customarily and wrongly described as the plastics pollution problem. Occasionally, there are bizarre benefits. In 1992 a Pacific storm caused 29 000 plastic bath toys to be washed overboard, resulting in a profusion of yellow ducks and green frogs and the like being washed ashore along the coast of Alaska in the ensuing months. 'The quack heard round the world' turned out to be a mine of information for oceanographic studies of the North Pacific.

Litter is the least apocalyptic of the disasters predicted by the more extreme environmental lobby. Sometimes it seems that every new success of plastics must perforce be capped by a new disaster scenario. Poisoning threatens on all sides, by water pollution, by incinerator gases, by migration into food, by widespread carcinogens. Problems there are indeed, some of them serious: but not much that should be beyond the reach of an ordered society, working with common sense and effective regulation. These weighty matters are addressed at length in the final chapter.

2. BRIGHT NEW WORLD
Innovation with Plastics

NOT ALL INNOVATIONS ARE NEW ONES

Twist-ties, crushable bottles, maintenance-free shock absorbers; there are so many familiar artifacts which could not exist without plastics. It would be a mistake however to give all the credit to our exciting new technology. The materials are new, but the methods are often reworkings of old concepts. Some have their origins in practices developed hundreds or even thousands of years ago, by practical people with practical needs, unblessed by modern science.

For many centuries horn was shaped and cut into a variety of articles, but particularly combs. Pressed into thin sheet it was used as a substitute for expensive glass panes. Modern techniques of shaping sheet by heat and pressure, and of compression moulding ground and blended powder, both derive from very old horn practices. Shaping processes for other natural thermoplastics like amber and tortoiseshell can be traced back to Greek and Roman times. In the late eighteenth century, these techniques were broadened to produce shaped articles in papier maché; this novel material was later enhanced by the 3000 year old multi-layering techniques of Chinese lacquer, to create rigid and waterproof surfaces for decorative boxes and trays.

Natural rubber and gutta-percha, both derived from secretions from trees, had been used as flexible coatings and shapings for centuries in their tropical heartlands (Columbus apparently encountered rubber balls in Haiti). Both materials were harnessed to the demands of nineteenth century industry, mainly for electrical insulation using the extrusion process, which in turn had its roots in the Italian pasta industry.

Even some of the developments we associate with the cutting edge of technology were in some sense anticipated by the ancients. Structural composites, for instance, can trace their antecedents back to brick making, wattle and daub building, Welsh coracles and Pacific reed boats, and possibly even Assyrian war chariots. It must have been apparent to the Ancients as it is to us that the natural world is extraordinarily rich in composites. Wood in all its forms is a natural composite. Skin and bone in combination form a macro composite; but each is itself a microcomposite, made up of fibres in a continuous matrix.

The range of effects achieved with polymers has been greatly extended by introducing particulate fillers, gas bubbles and reinforcing fibres. The properties of the continuous matrix, the polymer, then become only one of several factors determining the total effect.

There is another example of a new process with ancient origins: the production of hollow shapes by fusible core injection moulding. This has its roots in the lost wax process known to the ancient Egyptians. An old concept, but a massive input of innovative design and process engineering was needed to ensure that automomotive intake manifolds could finally go into volume production in the 1990s.

When an application is developed in plastics, success depends upon a harmonious combination of three factors: polymer, design and processing. So it is with all innovations: all three factors play a part. Usually, however, one of them is dominant.

POLYMER BASED INNOVATIONS

Fibres, Hinges and Bottles: Making Crystallinity Work

The ability of semi-crystalline thermoplastics to be drawn out in the melt and stretched into high strength fibres was noted during the early research into polyamides (nylons). This concept of strengthening a polymer by stretching, to orient the crystalline parts of its structure, was the basis of Du Pont's development of the nylon fibre industry in the late 1930s. The orientation process worked for high diameter monofilaments as well as fine diameter multifilament yarn for textiles. Thus it was that some of the early successes of nylon as a plastic (rather than a textile fibre) were achieved in monofilaments, not only in mundane articles like brushes, but also in demanding items like fishing lines, racquet strings and lightweight zip fasteners.

Polypropylene, following on some 25 years later, benefited from nylon monofilamant technology. Having weaker lateral forces between molecules, polypropylene can be more completely oriented. Very strong fibres result, in such uses as carpets, twine and high performance rope. The near-perfect orientation possible in polypropylene makes it outstandingly successful in strong integral hinges. The hinge is a band of highly aligned molecules, induced by stretching a moulded-in membrane. The integral hinge has been exploited in a huge variety of containers, fasteners and packaging devices, and for a while was used in accelerator pedals. The folding chopping board is a recent innovation.

The parts of a moulding that have experienced extreme orientation become more opaque: from this comes another application of the crystallinity of polypropylene: printing by stress whitening.

Thermoplastic polyester, in the form of PET, has developed along parallel lines to nylon as a textile fibre. These two polymers support a well-established textile industry. One market however belongs uniquely to PET: bottles for carbonated drinks. This application swept to stardom in the 1970s, through a

Fig. 2.1 Chopping board with polypropylene hinges (courtesy William Levene Ltd)

fortunate combination of clarity, low permeability, and (compared with glass), toughness and lightness. Whereas the textile applications exploit its crystallinity, the PET bottle market is based on the fact that under the right conditions the crystallinity of PET can be totally suppressed. It is injection moulded and then very rapidly cooled, into a completely amorphous and clear preform before the blow moulding stage.

Plastic Films

Plastic film has arguably been the greatest single agent for social change. It is certainly the most extensive application area. Packaging is of course the most visible end-use, but the benefits to civil engineering, to building and to agriculture are incalculable. The basis of this vast new industry, less than 50 years old, is the molecular versatility of thermoplastics. It is this which provides the opportunities for 'tailoring' in terms of melt strength and crystallinity, offering control over transparency, permeability and mechanical performance.

The two concepts of film and crystallinity come together very successfully with oriented film. The whole audio and video recording industries are based on it. The strength, lightness and dimensional precision of oriented PET tape provides the perfect base for the iron oxide magnetising medium. This was a long-sought solution: no other material could compare.

Plastic Foams

Foam technology started with rubber, and now with plastics it has generated a range of remarkable effects. There are choices between rigid and flexible

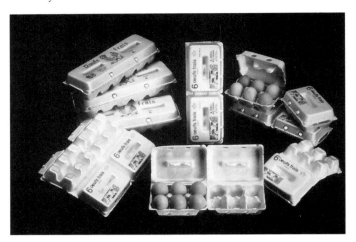

Fig. 2.2 Expanded polystyrene egg packaging

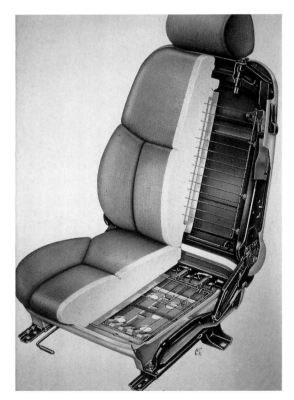

Fig. 2.3 The 1989 Ford Fiesta 'Unit Seat': foam shaping and fabric covering in a single operation (courtesy ICI)

foams, and between open and closed cell foams of quite different compressibility, permeability and insulation performance. This all arises from exploiting the chemistry and physics of long chain molecules, to introduce well dispersed gas bubbles into a thermoplastic melt before it solidifies, (or, in the case of polyurethane, while the monomers are reacting).

Expanded polystyrene (EPS) can be shaped into extremely light-weight foams, containing as much as 98% air. This has created entirely new possibilities for domestic thermal insulation and in the packaging of fragile articles.

Polyurethane in the form of closed cell rigid foam is central to the world's refrigeration industry, (and is being adapted to the impending ban on CFCs). New sophistication in open cell technology enables foams to be made with controlled variations in flexibility. This has greatly simplified car seat manufacture.

The necessary balance of resilience and support in a car seat can now be achieved with polyurethane foam on a rigid (plastic) base, instead of the traditional and expensively elaborate lattice sprung designs. A thin backing of polyurethane foam provides new effects in 'soft feel' car interiors and garment textiles. The huge sport shoe industry (with or without designer labels!) has grown out of the availability, along with synthetic fibre textiles, of a range of plastic foams. LDPE foams are now in use for widely differing applications. One of the most impressive is the use of closed cell foams to make massive but lightweight mooring buoys, for use in the Middle East.

Fig. 2.4 Mooring buoys in LDPE foam (courtesy Zotefoam Ltd and Hippo Marine)

Low Friction Effects

The concept of using resilient non-metallic materials in mechanisms with moving parts is much older than the modern plastics era: leather, horn and wood all saw service in gear systems in traditional industries. Evaluating the early synthetics in this kind of application followed quite naturally. It was apparent that some polymers exhibited low coefficients of friction and a natural lubricity. Nylon, always containing a little water, was the outstanding example, to be joined by acetals when they appeared in the 1960s. Since that time we have enjoyed a dramatic reduction in vehicle maintenance requirements: much of it is due to the introduction of nylon and acetal into such items as bushes and guides in assemblies like suspension arms, telescopic dampers and shock absorbers.

PTFE exhibits remarkably low coefficients of friction, but it is usually most effective when dispersed in other materials. Nowadays it is quite common to include more than one internal lubricant to achieve a particular performance from moving parts. Excellent results have come from specialist formulations, such as those pioneered by the LNP Company. Typically, these are based on nylon 66 or 6, with additions of PTFE, and graphite or molybdenum disulphide (having a platelet structure they reduce sliding friction), and also a silicone product which progressively releases oily material to the surface.

Another variant is shown in Fig. 2.5, a Mercedes ignition starter switch. The plastic insert is designed to give frictionless rotation, without the adverse effects of grease in make-and break circuitry. The base material is again nylon 66, with PTFE and aramid fibre, (less anisotropic than glass).

Today there is a vast range of appliances dependent on gears, bearings, cams, sliders and the like, quiet running and maintenance free. They have penetrated every corner of life, and most of them are taken completely for

Fig. 2.5 Ignition starter switch (courtesy LNP).

Fig. 2.6 Acetal pressure switches (courtesy Du Pont)

granted. Many of them, like nylon curtain runners and acetal pressure switches, use unmodified polymer.

The limitations of plastics in moving part applications are centred on temperature. Abrasion increases with temperature, because the surface becomes softer and thermal expansion reduces clearances, As surface debris appears, a runaway situation can develop. Polymers which are more heat resistant and with lower thermal expansion can out-perform nylon in these situations. Needle bearing cages in automotive gear boxes are a good example: glass reinforced nylon 66 is used in INA gear boxes, but where the ambient temperature exceeds about 120 °C, glass reinforced PES (polyethersulphone) is more durable.

Fig. 2.7 Automotive needle bearing cages in reinforced PES

Specialised engineering materials are not the only polymers to exploit low friction effects. The artificial ski slope is a spectacular example, usually made up of repeating units of LDPE. It is designed to give optimum contact and form stability, and takes advantage of the low friction of the material, especially when wet.

DESIGN-BASED INNOVATIONS

FORM AND FUNCTION ON THE DOMESTIC SCENE

The early records of plastics replacing metals were blemished by applications which went wrong. A common fault was to copy the shape of the metal article, instead of trying to reproduce its function by designing around the properties of the new material. Disasters based on this confusion between form and function are rare nowadays, but the possibility is always there!

Some of the best examples of this new understanding can be found in a very mundane domestic setting: coat hangers, for instance, and handles and hooks of all description. Curtain fittings tell a similar story. The brass rails have given way to extruded PVC, and ingeniously designed nylon runners and fixings have replaced steel wire and brass.

A search of kitchen cupboards and especially medicine chests reveals a rich display of design innovation. Pouring devices, child-proof stoppers with floating screw threads, tamper evident screw caps and user-friendly handles are all examples of focussed design exploiting the versatility of plastics.

From the earliest days, the toy market has been a happy hunting ground

Fig. 2.8 Tamper evident screw cap (courtesy Moss Plastic Parts Ltd)

for plastics. It has also been a difficult one to regulate, and although it abounds in good design at the quality end of the market it still features many stark examples of how not to use plastics. One conspicuous success has been the 'Lego' range, which depends on the good looks, colouring versatility and close dimensional tolerances of ABS. Bulkier toys make good use of injection moulded polypropylene and polyethylene, exploiting their toughness and stiffness, allied to bright colours and a snap-fit capability.

Design-based innovation is very widespread in garden equipment. There has been imaginative redesign of articles like rakes, watering cans and wheelbarrows into functional but unfamiliar shapes (such as the spherical barrow wheel). The garden kneeler-seat asssembled by force-fitting three blow mouldings in polypropylene or HDPE is a good 'harmonious' example in plastics: it owes much to the material properties and to the blow moulding process, but the decisive contribution is the design concept.

Design in the Automotive Industry

The automotive industry has become heavily dependent on plastics. Here the dominant criteria are weight reduction, mechanical and chemical performance, and (most of all) cost saving by component consolidation and simplified assembly. Three examples are relevant here: sound material selection and good processing are vital in all of them, but the key feature is design:

- Intricate designs of trim clips and fasteners are very widely used for assembling interior and exterior trim. These are injection moulded, usually in nylon, but also in polypropylene and acetal. Some are intended for repeated assembly, and others for permanent locking, making use of a sacrificial membrane.

Fig. 2.9 Fuel tank (courtesy BASF)

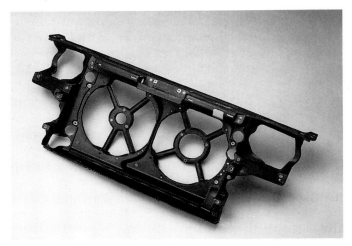

Fig. 2.10 Volkswagen front end frame in GMT-PP (courtesy BASF)

- Traditional fuel tanks welded from sheet steel are being progressively replaced by blow mouldings: the material is essentially HDPE of very high molecular weight, modified in a variety of ways to reduce its permeability to petrol. One crucial advantage is the behaviour in a fire situation; the plastic tank softens and burns, but does not explode. The other advantage is that by clever design, using the sophisticated controls of modern blow moulding, it is possible to produce tanks in almost any shape, to fit the space available. Before plastics, this would have been inconceivable.
- Front end frames, enclosing the front of the engine compartment and supporting several components, originated with the Peugeot VERA concept car of 1983. Different glass fibre reinforced materials have been used in recent models; injection moulded short fibre reinforced nylon and polypropylene, glass reinforced polypropylene in the form of GMT sheet, and thermosetting polyester as sheet moulding compound (SMC). (See chapter 6 for explanations of these products.) This particular application is an excellent example of how design, material and process work harmoniously together to yield different solutions appropriate to each set of circumstances. All of the designs are based on the consolidation principle, that a single plastics processing stage replaces a large number of metal fabrications.

PROCESS-BASED INNOVATIONS

Blow Moulding

Blow moulding provides many examples of process-led innovation. It is a process which is ideal for thermoplastics. Glass is the only other familiar medium allowing hollow shapes to be formed from the melt: however it is much less versatile, and in common use it is rigid and brittle. The domestic scene features a huge range of bottles for packaging liquids, mostly in PVC, LDPE and HDPE and (especially for carbonated drinks) PET. Shapes are suited to specific tasks, often with built-in handles and pouring facilities, with precise control of wall thickness to ensure the form stability of large containers and control the 'squeezability' of small bottles. The benefits of the clear PET bottle for carbonated drinks are well known. Less documented is the simple pleasure for the consumer of dispensing tomato ketchup from a squeeze bottle, without the hazards of emptying the traditional glass bottle.

Complex irregular forms such as milk bottles or fuel tanks need tight control of the viscosity of the plastic melt, and of the shape and wall thickness of the extruded parison offered to the mould before blowing. Computerised process control is essential.

Other Extrusion Processes

One of the most exciting innovations in extrusion technology is the 'Netlon' process. This is based on the elegant idea of using counter-rotating die-heads, in such a way that the melt threads intersect and fuse together before they can solidify. Many different materials are extruded to produce netting to all manner of strength specifications, in packaging, gardening and building applications.

Polymer formulations suitable for film were available long before the necessarily sophisticated production technology. Tubular LDPE film translated relatively easily into protective wrapping for the dry cleaning business. However, transforming it into such marvels as refuse sacks with draw strings, gussetted and printed upon, double-skinned in two colours and sold on a roll with tear-off perforations, required a wealth of imagination and a huge investment in machinery design and production engineering. More recently,

Fig. 2.11 Polypropylene netting (courtesy Netlon Ltd)

polyethylene bubble-pack film has given another illustration of a central anomaly, that to manufacture simple and expendable everyday articles to exacting quality standards at very high production rates calls for the very best in production technology.

The yogurt pot trade (and much more besides) was also the end product of much new technology: flat film extrusion, new printing techniques and high speed vacuum forming, filling and sealing. Extrusion and forming of polystyrene and polypropylene proved much more efficient and reliable than alternatives in glass or treated card.

Injection Moulding

Injection moulding is now an extremely important process, attracting huge investment in development effort. Innovations in design are intimately mingled with innovations in the process, and often they are mutually dependent. The new features of injection moulding are summarised in chapter 5, but two are worth singling out as having initiated new applications:

- The controlled use of internal voids in mouldings, either as a foam or as large continuous bubbles, has in turn led to structural foam moulding and gas assisted injection moulding in its different forms. Products have benefited from new arts like the ability to form internal channels during moulding, or the concept of making an article stiffer while actually reducing its weight (see Fig. 5.5).
- Based on ancient technology it may have been, but many years of effort in many countries was required to develop the fusible core process, extending

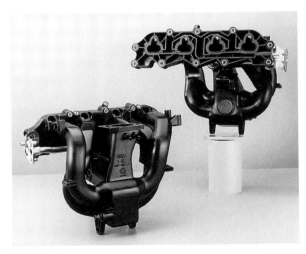

Fig. 2.12 Intake manifold with exhaust gas recirculation, moulded by fusible core process in reinforced nylon 66 for Ford 1.6 Zeta engine (courtesy Du Pont)

the range of the injection process to hollow mouldings. Competition grade tennis racquets were the first commercial success, in carbon fibre reinforced nylon. Intake manifolds for large and medium automotive engines have now been established, first in reinforced epoxy but latterly in glass fibre reinforced nylon 66.

3. WHY CHANGE TO PLASTICS?
The Pros and Cons of Replacement

Chapter 2 signalled the innovations derived from plastics; the novel effects which our grandparents never knew. In most applications however, plastics are simply replacing a traditional material. There can be many possible reasons for the change. Some of them are obvious; others much less so. The choice is usually based either on cost benefits or performance benefits (although the two cannot always be neatly separated).

COST FACTORS

RAW MATERIAL COSTS

The cost of any industrial raw material is built up from factors such as these:
- The availability of its original natural materials.
- The complexity of the process of extraction and purification.
- The energy this process consumes.
- Packaging, storage and transportation.

Globally, these factors are often more favourable to plastics than might be expected. What is probably the most high-profile factor, the price of crude oil, is not the dominant influence on plastics prices. The unpredictable laws of supply and demand still exert a more potent effect.

Comparing the energy involved in the extraction of materials usually gives a result favourable to plastics (especially opposite aluminium).

Commodity polymers apart, the direct material costs in terms of price per kilogram are not very favourable to plastics. Translating into price per unit volume gives a more logical basis for comparison, especially for components whose dimensions are largely fixed. However, material cost is only the beginning.

COMPONENT COST

In any complex manufacturing operation, accurate cost accounting is crucial. When there are many components and large volume production, the effect of quite small costing errors can be decisive. The automobile industry, assem-

Table 3.1 Relative energy expended in extracting equal volumes of materials

LDPE	1.6
HDPE	1.8
Polypropylene	1.8
PVC	1.9
Polystyrene	2.0
Steel	4.6
Copper	10.8
Aluminium	15

bling hundreds of thousands of units each comprising thousands of components, is the ultimate costing challenge.

In this extreme case, the automobile, the critical parameter is the 'piece part' cost, to which many things contribute:

- Raw material cost
- Tooling cost
- Total number of parts
- Direct conversion cost per part
- Cost of sub-assembly
- Assembly line costs

There is the ever-present problem of balancing the equations which relate total number of parts, number of tools, production rate, tool life, etc, against a background of uncertain sales figures. Nowadays with the rigours of 'Just-in-Time' assembly there are additional costs incurred, in ensuring that the rate of production and distribution at each stage matches the demands of the next stage. The Just-in-Time culture, involving as it does components from many sources, very often in different countries, is also highly vulnerable to extraneous influences in the shape of natural disasters and volatile industrial relations.

It will be appreciated that one decisive factor in changing to plastics is 'the numbers game', posing the perennial question: Do the intended production numbers justify the up-front costs? If changing to plastics implies totally different fabrication methods (which is often the case!), then the up-front costs are not limited to tooling but must also include capital costs of presses and assembly line modifications.

Consolidation

Time and again, the factor which tilts the balance in favour of plastics is consolidation. This is the concept of replacing several discrete components and fabrication stages by a single processing operation.

Examples of consolidation are all around us. The moulded-in handle is

Fig. 3.1 Washing machine drums in steel and polypropylene

very evident in the kitchen: blow-moulded in a variety of disposable bottles for drinks and detergents, and injection moulded for appliances like electric kettles. Changing the outer drums of front-loading washing machines to plastics has replaced a stove enamelled 20-stage fabrication in sheet steel by a single plastic moulding in reinforced polypropylene. Only the steel bush has been retained from the original.

From the motor industry, the integrated 'front ends' now in common use are a spectacular success for consolidation (Fig. 2.5).

Cost Benefits in Service

There are many 'follow-on' benefits which can be recognised in a change to plastics. The fact that a calculable cost advantage shows up somewhere along the supply chain is usually enough to initiate the change. Examples show how varied these advantages can be:

- Weight reduction: This is rarely an overriding consideration, except in special cases like the aircraft industry. As a secondary factor, however, it is assuming enormous significance for ecological reasons. Weight reduction in motor vehicles is very important because of lower fuel costs, and also because of reduced emission of toxic gases. It is also a big factor in bulk transportation costs. It has helped to establish PET as a replacement for glass bottles in the carbonated drinks industry.
- Introducing nylon gears into agricultural machinery, handling potentially corrosive fertilisers under inadequate maintenance schedules, can greatly increase its life.
- The reduced maintenance required by the modern car mean lower running

costs for car owners. The main agent for change has been the exploitation of the lubricity, durability and load-spreading capability of engineering plastics, mainly nylons and acetal, in a wide variety of moving parts.
- Reduced assembly costs and elimination of breakages have contributed to LDPE sheet replacing glass in a host of market gardening activities.

PERFORMANCE FACTORS

Freedom of Shape and Form

Melt forming gives the designer the possibility of freedom of shape, far more than in fabricating from standard forms like sheet and rod. Injection moulding allows all the design freedom of die casting, but with far fewer finishing operations. Much the same applies to in-mould polymerisation processes like reaction injection moulding and resin transfer moulding. Likewise blow moulding and rotational moulding can provide complex hollow shapes in a single operation.

Styling freedom has been the decisive influence in many kitchen appliances, and in automotive components like door-mounted mirrors and rear spoilers. The front and rear ends of today's cars all make good use of this styling freedom, whilst meeting a challenging list of demands like energy absorption, dimensional stability, paintability, and resistance to chemicals, abrasion, temperature and sunlight.

Rational Design Capability

Precise control over both surfaces of an article, such as injection moulding provides, means that strength and stiffness can be controlled by ribbing and wall thickness. Deliberate changes in the wall thickness allow such refinements as pre-determined fracture points (eg nylon in single-use trim clips) and integral hinges (eg polypropylene in coffee maker lids).

Single surface forming, as provided by blow moulding and rotational moulding, is less precise; but this is irrelevant in applications like bulk containers (in polypropylene) and automotive fuel tanks (in HDPE), all meeting exacting specifications for strength and form stability.

In the same way the design flexibility of plastics is used to good effect in extruded products. Accurate control over wall thickness and diameter enables pipe and tubing to be made consistently to demanding building and engineering specifications. Examples are automotive fuel lines in nylon (PA 11 and PA 12), water supply pipes in LDPE, and, most importantly, vast numbers of public utility pipes and conduits in PVC. These applications can use another design feature to good effect: 'built-in' colour coding.

COLOURABILITY

Colourability is an important aspect of design freedom. It was the development of plastics which were both mouldable and colourable, in particular urea-formaldehyde (UF) and cellulose acetate (CA), that opened up an extensive 'domestic artifacts' market in the 1930s. Nowadays it is the self-coloured thermoplastics, especially polypropylene, which are to be seen everywhere in articles for kitchen and garden, and especially in toys. The success of PP in the toy trade may be due as much to its colourability as to its fortunate balance of rigidity and resilience.

Colour coding is a concept which links a wide selection of plastics applications. Coloured polypropylene boxes feature in office storage systems and tote boxes for workshops and assembly lines. Electrical circuitry in cars and appliances is colour coded with the aid of nylon connectors and PVC cable insulation. Colour becomes a safety feature in all manner of appliances, as in the use of red quick-release buttons in car seat belts.

Consistent 'solid' colouring in plastic articles depends on uniform dispersion of the pigments. This owes much to modern compounding equipment, particularly screw injection machines and extruders. It was more difficult in the early days, with materials like celluloid sheet and the translucent compression moulding powders which followed. Designers made a virtue of necessity, with artificial tortoiseshell and 'marbled' effects becoming popular. Simulated wood effects (always dear to designers of British car interiors!) were popular in the 1930s with wood-filled PF mouldings. Similar mottled decorative effects could also be achieved with thermoplastics on the early plunger injection machines, by incomplete mixing of different colours and viscosities.

The solid colouring of pigmented plastics sometimes allows an expensive painting stage to be eliminated: electrical hand tools are a good example, with coloured reinforced polypropylene replacing painted diecast aluminium. It has to be appreciated however that even a high gloss plastic shaping can never be smoother than the steel surface which moulded it. Furthermore a self-coloured plastic surface will deteriorate much faster in use than one which has been painted and oven-cured. Hence for a 'hybrid' car body with a mix of metal and plastic body panels it is imperative that all the panels are painted, (preferably with the same paint system), unless the designer deliberately selects contrasting colours or textures.

Precise colour matching of plastics demands a much higher order of expertise than merely colouring. Matching between a plastic moulding and (say) a painted surface or a textile surface can be extremely difficult, simply because the texture is different. It is even difficult to colour match between different plastics; and furthermore what works for one set of lighting conditions may not work for another. Designers are well advised to look for 'sympathetic' colours rather than total uniformity.

Safety

Plastics make a direct contribution to safety in three clearly defined respects: energy absorption; insulation (electrical and thermal) and lubricity.

The impact strength that results from energy absorption benefits us at the most basic level, with 'bounceable' bottles and shatter-proof sheeting. In all applications, the energy absorbing and cushioning benefits need to be accompanied by good design. There are many such examples from the motor industry: passenger compartments with recessed door handles (Fig. 3.2), steering wheels which are dished and covered in self-skinning polyurethane (see Fig. 5.8), and resilient, cushioned fascia panels.

Other vital safety features include brake fluid reservoirs in polypropylene; nylon fuel lines; and of course the HDPE fuel tank. The increasingly sophisticated passenger restraint systems of seat belts and air bags are composed almost entirely of synthetic polymers and textiles. Protective clothing also relies on the energy absorbing properties of plastics. The range extends from hard hats, safety glasses and all the motor cyclist's equipment to specialist products like 'flak jackets'.

The electrical supply industry could not function without plastics. Circuitry depends on cables insulated with PVC, and connectors, terminal blocks and wall sockets mostly in PF, UF and nylon. Metal housings for electrical hand tools were often lethal when incorrectly wired: since the 1960s tools like hand drills have been redesigned with double insulation, made possible by non-conducting casings moulded initially from nylon and latterly from reinforced polypropylene.

The benefits of thermal insulation are less dramatic, but 'cool touch' toasters and kettles make life in the kitchen more comfortable. Expanded polystyrene cups make hot beverages easier to carry (but no easier to drink!).

The safety benefits of lubricity are not easily quantified, but it can be said that the effects of poor lubrication are likely to be less severe with plastics

Fig. 3.2 Component of recessed car door handle in glass fibre reinforced nylon

than with metals. The fact that tolerances are less critical also helps, particularly in gear trains. Plastics are more 'forgiving'!

There is another characteristic of plastics which overall makes a bigger contribution to safety, because it underpins so many end uses. This is the provision of chemically inert and bio-compatible surfaces, for promoting hygiene and preventing toxic contamination. Married to their versatility in forming and shaping, this means that plastics are catering for civilisation's most basic needs. Examples like pipes and containers for potable water, silos and storage bins for commodity foods, and a range of medical uses from hip joints to blood bags are witness to this claim.

Corrosion

Corrosion in metals can have devastating consequences for Man and all his Works. It can hurt at all levels, from the premature demise of a lawn mower to a catastrophic train crash.. Battles have been lost because of corrosion. (At Isandlwana in 1879 the British Army was unable to open its ammunition boxes, their steel clasps rusted after the long sea voyage: the result was the most spectacular massacre of the Zulu Wars.) Nowadays enormous resources are expended in coating metals or changing the chemistry of their surfaces. Car bodies are given expensive multi-layer protection by complex new technology. And the painting of the Forth Bridge continues ...

The trouble and expense of anti-corrosion measures are a major incentive for the change to plastics. A familiar homely example is the injection moulded polypropylene paint pot, which has ensured that rust-stained pots are a thing of the past.

Acoustics

One of the many benefits of plastic foam is sound insulation. It is taken for granted nowadays that automobiles are quiet running, at least insofar as their passengers are concerned. The principal agent for this change is polyurethane foam, which is used to insulate the floor pan, the roof, the hood, the engine bulkhead and the doors. Normally, the flexible foam is sandwiched between the panel and a 'massback', in the form of a layer of PVC or LDPE with its density increased by loading with heavy inorganic fillers. (Interestingly, this is one application in which the light weight of plastics is a disadvantage.) In some recent models, polyurethane has been polymerised directly within the hollow sections of the body structure.

The rattles and roars of early vehicles are but a memory, but their absence reveals all manner of lesser squeaks, hummings and buzzings. Unsurprisingly, the squeaks are attributed to plastics: specifically to rubbing between different parts, with dimensional variations being blamed for day-to-day variations. Currently the matter is receiving much dedicated attention, with international discussion groups in place. (Once again, progress exposes new problems, the goalposts are moved, and new research targets are funded.)

Rattles have in fact disappeared, unmourned, from many walks of life. Gears in nylon and acetal ensure quiet running in all manner of mechanical devices (but this is not new; horn was used in clock escapements long ago, for quieter operation). The traditional British early morning delivery of milk in glass bottles is a very much quieter affair, blessed with the vibration damping of polypropylene, than it ever was with galvanised steel crates.

PLASTICS AS SUBSTITUTES: HOW GOOD ARE THEY?

THE GENERAL CASE

Against traditional materials plastics score consistently because of three effects:

- Their weight saving potential
- The design freedom bestowed by the various moulding processes
- The cost benefits of scale provided by these processes.

Nevertheless, beyond these three effects, there may well be some negative factors. It is useful to set up a basic check list of 'pros and cons' for each type of material replacement.

GLASS REPLACEMENT

Before the introduction of cheap plate glass, thin sheets of horn were widely used. This cost advantage over glass is unlikely to reappear for modern thermoplastics. The spectacular success of acrylic sheet for aircraft cockpit covers in the Second World War led many to predict a similar takeover in the automotive industry. However, this has not proved to be viable, although PMMA sheet is well established in caravans and even in coaches. Weight saving will always provide an incentive for car windows, but the superiority of toughened or laminated glass as a scratch-resistant windscreen material (sandwiching a layer of polyvinyl butyral) is unchallenged. However, glass/polycarbonate laminates are now under investigation for side and rear windows. Now that legal objections have been overcome, glass is being replaced in car headlamps; the dominant factor here is the design freedom of moulded polycarbonate (enhanced by a scratch resistant coating).

The long-standing role of glass as the universal bottle for each and every liquid has now been usurped. Plastic bottles have largely taken over in the kitchen: the reasons are a mixture of cost, convenience, toughness, shape freedom, and 'squeezability' – an irresistible attraction for viscous liquids.

Fig. 3.3 Polycarbonate headlamp lens

Most of the food applications that still use glass do so for reasons of 'industry culture' (as in the wine trade) rather than the inadequacy of plastics.

Nevertheless the virtues of glass are real enough. The scratch resistance, dimensional stability, solvent resistance, durability, chemical purity and softening point are decisively better than plastics. Very often however these differences are not crucial, and then the issue is decided on the convenience factors of plastics, most particularly 'unbreakability' and pay-load. The extraordinary growth of PET for carbonated drinks bottles is the prime example.

Wood Replacement

The status of timber (or its derivatives) as basic, all-purpose structural raw materials is probably as secure as it ever was. For shaped articles, however, material selection is increasingly decided by the numbers game. Although wood working operations are now significantly automated, high volume production of shaped items such as door handles will always favour plastics. There is a considerable trade in replica furniture, using such materials as foamed reinforced polypropylene and polystyrene, in which it is possible to match the density and rigidity of wood fairly exactly. Some designers prefer to work in 'genuine' plastics, rather than the 'halfway house' of simulated wood, in the form of veneered woodchip. As the new culture of recycling plastics grows apace, new wood-substituting applications appear. Garden trellis, for instance, extruded from recycled polystyrene, keeps its looks far longer than 'the real thing' in wood.

The enormous popularity of unplasticised PVC window frames is based on their resistance to the rigours of atmospheric exposure (Fig. 3.4). Painted wooden frames can only be protected from the usual destructive agents (sun, rain, wind, oxygen, sea salt, acid fall-out, frost and temperature cycling) by continual and expensive maintenance. This, and the ecological virtues of

Fig. 3.4 Window frames in unplasticised PVC (courtesy BPF)

double glazing, should on all rational grounds secure this application for PVC. For non-structural parts such as soffits and other cladding, cellular PVC is well established, being much lighter without losing the weathering advantages.

We can be in no doubt that the 'quality' end of the wood market is secure, and probably always will be. The visual and tactile supremacy of grained and polished wood, shaped using its natural anisotropy, is unchallenged.

Challenges have come however in an unexpected quarter: sporting goods. Sporting equipment is noted not only for the punishing mechanical stresses it endures, but also for its power to siphon off surplus spending power from the affluent. Some of the first conspicuous successes for carbon fibre came from this sector: with fishing rods, vaulting poles and golf clubs in the form of thermosetting polymers reinforced with continuous carbon fibre. Even more startling was the 'top of the range' tennis racquet. The material was nylon reinforced with short carbon fibre, and the correct weight was achieved by injection moulding around a temporary fusible core, to give a hollow section.

Ceramics Replacement

As every archaeologist knows, the most durable of Man's artifacts are ceramic. Of some civilisations, little remains except a few fragments of pottery and glass. What this apparent obsession of ancient civilisations with pottery tells us, of course, is simply that nothing else has survived.

Drinking cups and storage pots for liquids were the basic application. Although horn, leather and dried fruit skins were also used, pottery was always the favourite, when the raw materials and the skills were available. As civilisation advanced, generating more sophisticated needs like dishes and plates, the superiority of glazed ceramics became evident. Plastics have made few inroads into the market for glazed china, established in Britain since the import of Dutch glazing techniques, late in the seventeenth century. Celluloid and later translucent thermosetting resins were used in cups, dishes, vases, etc, for their decorative possibilities; valued now as collectors' items, but functionally they could be called 'gimmicky'. Today, as far as kitchen crockery goes, plastics have done little more than take over the 'picnic' niche, with melamine-formaldehyde because of its light weight, robustness and scratch resistance. However, injection moulded PET is accepted as a material for microwave utensils, by virtue of its temperature resistance.

In the 1980s silica-filled acrylic compositions began to encroach upon the market for glazed domestic ceramics, with newly designed shapings for decorative coloured sinks and wash basins.

Plastics are now even making an impression on the strictly functional end of the market for glazed sanitary stoneware. Glazed sewage pipes were the means by which Henry Doulton began to civilise Britain's cities in the late nineteenth century: where sewage is involved, durability and impermeability are all-important. However, PVC is now accepted for catch-pots, junction chambers and all diameters of sewage pipes, against very demanding specifications. PVC piping is easier to install, it is far less likely to fracture in unstable subsoil, and its ability to cope with high pressure jet cleaning has now been established. For applications involving scouring, of course, ceramics are still secure, because of the extreme hardness of the glazed surface.

The garden has proved a more successful hunting ground for plastic pots than the kitchen: our garden centres are seemingly awash with a profusion of pots, mostly in polypropylene. Here the extra performance given by glazing is not needed, and hence the advantages of cost, weight and short-term convenience outweigh any consideration for future archaeologists.

Ceramics played an important part as insulators in the formative years of the electrical industry, and this was extended into the first volume-produced domestic electrical fittings. It soon became apparent that PF mouldings could be just as effective (at least at domestic voltages) and were much easier to manufacture. However, when a high ambient temperature is unavoidable, as in some lighting fittings, then ceramics provide much superior resistance to heat ageing.

Wherever ceramics are used because of their essential properties, their position is unassailable. In terms of heat resistance and flammability, abrasion resistance and durability, chemical inertness and dimensional stability, advanced ceramics out-perform plastics by a considerable margin. For many demanding industrial applications, ceramics are the natural choice. New processes, starting with fine powders of high purity and eradicating weak spots by greater densification are increasing their competitiveness, and opening up new horizons.

One example reaffirms how important it is to survey all the aspects, when selecting the best material for a job. Fish processing machinery needs cutting blades, which for reasons of contamination cannot be metallic. Blades were produced in nylon and acetal, both hard-wearing engineering plastics and cheaply fabricated: however they were abraded quite quickly. Replacing them by high strength zirconia, a much more expensive material, gave a more durable cutting system, with ultimate cost benefit.

Steel Replacement

Steel and plastics are so essentially different that it is perhaps surprising to find how often they claim the same applications. Outside this common ground, their limits of usage are determined (relatively and very generally) by some well-known characteristics.

Table 3.2 Steel and plastics: the weak spots

Deficiencies of Plastics	Deficiencies of Steel
Brittle at low temperature	Poor fatigue resistance
Poor high temperature rigidity	High density
Dimensional instability	Poor lubricity
Poor solvent resistance	Poor corrosion resistance
Poor scratch resistance	No integral colouring

(The logic of comparing weak spots, rather than strong ones, is that in practice material selection works largely by a process of elimination. It is the negative points which are decisive.)

Most replacements of steel by plastics have been achieved in cases where the temperature and chemical extremes do not apply, and when it has been possible to design out the plastics limitations in respect of rigidity and form stability. The need to thicken a section may be more than offset by the cost savings in eliminating machining. A good example is the automotive control pedal, either clutch or accelerator, where a fabricated U-section in steel is replaced by an injection moulding in glass reinforced nylon, in which the U-section is stiffened by heavy ribbing. There are advantages in strength, weight and cost.

The pedal example illustrates a general truth about changing to plastics: it is the complex shapes that are more likely to yield a cost saving. The moulding processes, especially injection, offer a single stage fabrication, often with the benefits of consolidation with other components. With simpler shapes, capable of being pressed from sheet steel, it is harder for plastics to

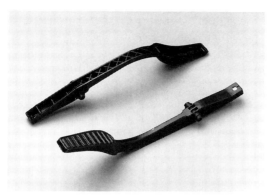

Fig. 3.5 Accelerator pedals for Ford Escort in glass fibre reinforced nylon

show a cost advantage. An automotive engine sump, for instance, can be produced in one or two straightforward pressing operations: there is no economic case for injection moulding. However, a pressing in a glass reinforced composite, not requiring expensive tooling, can be a much more attractive proposition.

It is important when extolling the virtues of plastics in metal replacement to remember one central fact: that is, the basic simplicity of sheet steel fabrication processes. They are usually highly capital-intensive and energy-intensive, but nowadays robots ensure minimal labour costs. All plastics fabrication processes involve at least one change of state (solid-liquid-solid), or two, or a chemical change, or both. Steel sheet for car bodies is shaped by pressure alone, without heat, albeit with subsequent stamping and spot-welding stages: and all at very high speed. Nevertheless the 1990s have seen increasing interest in direct replacement of certain steel body panels by plastic ones; even to the extent of the plastics being made conductive to undergo the electrocoat stage (see Fig. 4.7).

In the late 1990s the steel industry began to fight back. The ULSAB (ultra light steel automotive body) initiative has demonstrated that steel car bodies can be made much lighter, by reducing wasted flange width, for instance. In the car body area, the steel versus plastics contest will run and run. (And, as the focus shifts to small city cars, the goalposts will move yet again.)

Replacing Other Metals

There are many cases where long established metal applications disappeared very quickly when plastics arrived; or, rather, when a suitable plastic arrived along with a cheap and effective shaping process. Lead enjoyed unchallenged supremacy for perhaps 2000 years as conduits and pipes for the urban water supply. It is now being rapidly displaced by LD and HD polyethylene and PVC on health grounds.

Lead-based toy soldiers are another example, now almost forgotten. Lead was displaced in a very short space of time by injection moulded LDPE in the late 1950s. Oddly enough, the reasons were not health-connected, but based on the familiar arguments of cost, impact strength and colourability. Novel ideas like the 'Swappit' concept of exchangeable heads and limbs, so easily achieved in polyethylene, finally sealed the fate of the lead soldier (and enriched the young lives of the post-war generation ... !)

The 'profit and loss' performance balance of plastics against die cast metals (essentially aluminium and zinc alloys) is broadly the same as that against steel. The high temperature and crushability/ductility advantages of steel are absent from die castings, but so is the density penalty. In fabrication, of course, the die casting materials can offer nearly as much in design freedom as the mouldable plastics.

Improved grades of aluminium and zinc have narrowed the volume cost advantage of plastics, so that the metals are largely holding their own against the engineering plastics. However, in appearance-sensitive consumer applications, the need for finishing operations and painting tips the cost balance back in favour of engineering plastics. Sometimes property differences are decisive: the double insulation advantage means that for electrical hand tools, reinforced thermoplastics are now the materials of choice.

WHERE THE OLD AND THE NEW CO-EXIST

Although the process of displacement may seem never-ending, there are many examples of co-existence. Plastics articles have often appeared alongside the traditional materials, without actually supplanting them. The effect has rather been to enrich the choice available to an already over-indulged public. There are abundant examples in the department stores; zip fasteners,

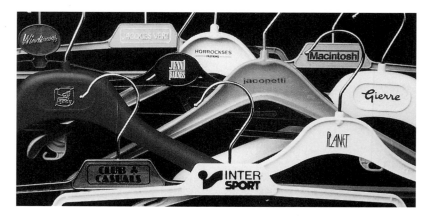

Fig. 3.6 Plastic coathangers (courtesy BPF)

electric kettles, and items of door and window furniture, for instance. Interestingly, the humble coat hanger is perhaps the most ecumenical item of all: wooden ones, steel wire ones, fabric-dressed ones, and dozens of designs in LDPE, polypropylene and polystyrene mingle happily in the family wardrobes. More often than not, it is the plastic hangers which are 'top of the range'; injection moulded with intricate multi-functional designs that would be impossible in traditional materials.

The DIY store shows the replacement process in action. The purchasing public with its own personal consumer trials generates the market forces which eventually ensure that only the best cost-effective designs survive. There, alongside numerous short-lived gimmicks and market failures, there are many items which exhibit the essential harmony between material, process and design. Increasingly too we can see illustrations of plastics and metals working together more effectively than either could alone.

4. SOMETHING FOR ALL OCCASIONS
The Nature and Diversity of Polymers

WHAT IS DIFFERENT ABOUT POLYMERS?

Since chapter 1 (or before!) we have been accustomed to thinking of polymers in terms of long chain molecules. Two things determine the behaviour of a polymer: firstly the chemical identity of the 'building blocks' which make up the molecules; and secondly the alignment, movement and interaction of the molecules themselves. It is this latter feature, this 'infra-structure' which makes polymers different from other materials.

CHARACTERISTICS DUE TO THE ORGANIC NATURE OF POLYMERS

- Densities are low, even when polymers are heavily filled with inorganic additives.
- Conductivity is very low, so that polymers are good thermal and electrical insulators.
- Generally polymers are resistant to inorganic agents like acids and alkalis, but sensitive to attack by organic solvents.
- All polymers are susceptible to surface oxidation and embrittlement, from exposure to heat or ultra-violet light.
- The basic carbon–carbon linkage means that polymers are inherently flammable, although some are much more resistant than others.

CHARACTERISTICS DUE TO THE MOLECULAR STRUCTURE OF POLYMERS

- There is always some structure in a polymer melt, so that its viscosity varies with temperature and strain rate.
- The detail of the 'infrastructure' is important. The properties of semi-crystalline thermoplastics depend on the rate at which the melt has cooled. Shock cooling can induce internal stress in the solid.
- Semi-crystalline polymers can be stretched, to order to align the crystalline

zones. This orientation dramatically enhances properties in the favoured direction. It is the basis of fibre and live hinge applications.
- Amorphous (ie non-crystalline) polymers are often inherently transparent.
- Polymers are generally more subject to dimensional change than other materials.
- Creep (ie, deformation continuing to increase under a constant load) is much more evident in polymers than in metals. Under load polymer molecules will move to accomodate the applied stress. For designers this can be an advantage; if ignored it can be disastrous.
- Because the molecules are capable of movement, the stiffness and strength of polymers are both time and temperature dependent. It is this fact which most differentiates polymers from metals and ceramics. Engineers sometimes find it upsetting.

THE PROBLEMS OF STIFFNESS AND STRENGTH

STIFFNESS

The stiffness of a material is characterised by a modulus, which is the ratio of the applied stress to the resulting strain. A material is said to have a high modulus if a load produces a small deformation, and a low modulus if the deformation is large. Essentially, metals have a constant modulus, largely unaffected by the temperature or the duration of the load: if you double the load, you get double the deformation. These are the characteristics of an elastic solid; reassuringly unchanging.

With polymers it is a different story. To start with, strain is not proportional to stress, except at very low stresses. The deformation is rather more than would be expected, and the greater the stress, the more disproportionate the resulting strain. In other words, the higher the stress, the lower the modulus appears to be. But the problem doesn't end there. If a load is maintained, the material creeps; so the apparent modulus diminishes with time. When the load is removed, the material begins to recover its shape. If it has stretched too far, it will have shown 'plastic' behaviour and yielded, irreversibly. Polymers however are not purely 'plastic'; if they were it would be impossible to design any load bearing components. They are in fact visco-elastic, meaning a mixture of elastic and plastic behaviour.

Temperature adds to the problem. Heating a polymer increases internal energy, and possible molecular movement. The modulus reduces even further: the material becomes less elastic and more plastic.

STRENGTH

Strength is also dependent on time and temperature. The concept is more complicated however, because the point of failure, the strength, can be defined in different ways. Ductile polymers fail by yielding (irreversible plastic deformation), while brittle ones fail by fracture. It is possible for a material to fail by both mechanisms, under different conditions. The essence of good design is to ensure that under the designated service conditions, neither kind of failure will occur.

Ductile failure is easy to understand; brittle failure is more complicated (and more catastrophic when it happens!), Essentially, what changes the behaviour from ductile to brittle is the inability of the molecular chains to accommodate an applied stress. This can be made to happen in three ways: by lowering the temperature, by increasing the speed of loading, and by allowing stress concentrations such as notches. All these factors can cause a material to perform below its potential by making it brittle. Chapter 8 tells in more detail how we can ensure that plastics do conform with expectations.

The behaviour of visco-elastic materials may seem alarming to those unused to them. However, once the properties of a particular material have been established (mainly by laboratory creep measurements), its behaviour can be predicted very accurately. Materials can be characterised by a 'design stress'. This is best defined as the stress at which the maximum strain acceptable in the application is reached. Using this 'maximum strain' criterion, ductile and brittle materials can be compared directly.

CLASSIFICATION OF POLYMERS

THE CATEGORIES

What has been said so far is general, in that most of it applies to most types of polymer. From here on, we must sharpen the focus by examining the different categories of polymer.

Firstly, the prime distinction is between thermoplastics and thermosets. Secondly, within the thermoplastics group it is useful to distinguish between crystalline and amorphous materials. Thirdly, elastomers can be separated from the rest of the plastics family.

Fourthly, there are a great many different chemical species of polymer, each carrying the 'trade marks' of the original monomer 'building bricks'. The differences between these chemical types can be crucial.

THERMOPLASTICS AND THERMOSETS

There is a fundamental difference between these two classes of polymer. Thermoplastics are formed by joining together repeated units of monomer into long chains. Heating a thermoplastic loosens up the chain molecules so that they can move independently, in a melt. As the melt cools, movement between molecules subsides. The melting and freezing cycle can be repeated indefinitely. (In practice there is a steady deterioration at each cycle, through contamination, oxidation, or breaking of molecular chains.) If heating is continued beyond what is necessary for processing, the melt becomes more fluid and eventually the polymer decomposes into liquid and gaseous products.

Thermosetting polymers are formed by the reaction between two monomers, with cross linking. Once formed by heat they cannot be reshaped. Shaping and forming must therefore be effected before the cross-linking is complete. The giant molecules which result are very stable and undergo very little movement under thermal and mechanical stress.

High temperature performance of thermosets is crucial. They will lose some volatile material and eventually char, but they will not melt and are much more difficult to ignite. These characteristics explain the continued popularity of PF and UF for domestic electrical fittings, saucepan handles and ashtrays. Other characteristics, unmatched by thermoplastics, are hardness, scratch resistance and chemical inertness. Hence UF and MF (melamine formaldehyde) are the favoured materials for tableware, decorative laminates and toilet seats.

In other application areas the differences are not so well defined. Thermoplastics in general exhibit better flexural, impact and fatigue behaviour, while thermosets tend to have better compressive strength and resistance to heat and abrasion, with significantly better dimensional stability. The decisive differences between the two classes are in the processing methods, with their associated cost and volume implications. Very large forms like boat hulls call for a thermosetting material and low-cost, low pressure tooling, whereas high speed moulding of small components in multi-cavity tools is largely the preserve of thermoplastics. However, between these extremes the issue is less easily decided. The scale of production and the up-front tooling costs are very often the dominant decision factors. Thermoplastics are generally more versatile for the various moulding activities and are enjoying a much higher growth rate. The enormous extrusion industry, which sustains most of the packaging and building industries, is of course entirely thermoplastic.

CHEMICAL TYPES

The chemical structure of polymers is all-important in determining the behaviour of the raw materials of plastics. The range is too complex and altogether too vast to present a compact and meaningful summary; but there are some useful guidelines.

Most thermoplastics have a molecular backbone of linked carbon–carbon atoms. This is true of the bulk polymers like polyethylene; it also fits an extreme case like PTFE. In PTFE the chains are not only very long, but also extremely compact: they pack very closely together, resulting in high density,

Fig. 4.1 Applications of PTFE

high crystallinity and non-stick properties. The resistance to chemicals and to high temperature is superlative, but PTFE cannot be melted and must be processed by sintering.

The polyolefines (primarily LDPE, HDPE and polypropylene) are semi-crystalline with the 'pure' carbon–carbon backbone, but without 'exotica' like fluorine. They differ significantly in detail, but share the common characteristics of excellent resistance to acids and inorganic chemicals and very low water absorption. The battery in Fig. 4.2 exemplifies two characteristics of polypropylene: its acid resistance and its integral hinge capability.

PVC is a special case, by virtue of the chlorine atom in the monomer molecule: in fact it contains 56.7% chlorine by weight. (This statistic has generated much ill-founded unpopularity: more of this in chapter 9). PVC is a very versatile low-cost polymer. Its great strengths are its excellent weathering and flammability performance; but its useful temperature range is limited at both the high and low ends.

Significantly different properties result when other elements are introduced into the carbon–carbon 'backbone'; eg nitrogen in nylon and oxygen in acetal. Both these classes of material have excellent mechanical properties, good resistance to solvents and organic chemicals in general, and useful 'self-lubricating' character. They are the quintessential 'engineering plastics'. The petrol filter assembly in Fig. 4.3 exemplifies the chemical resistance of nylon, in a thermally and mechanically exacting environment.

Throughout the 1960s and 1970s there was a proliferation of new high performance engineering plastics. Structurally, most of them are made up of linked 'benzene rings' instead of straight chains. These impart much greater resistance to heat, fire and chemicals, and also classify the polymers as 'aromatic'. The individuality (and the nomenclature) of these materials derives from the particular chemical groups forming the links between the

Fig. 4.2 Polypropylene battery

rings; hence polyethersulphone (PES) and polyphenylene sulphide (PPS), etc. It will be apparent from Table 4.1 that there is a relationship between cost and temperature capability: in fact the cost of building in extra short- and long-term high temperature performance can be considerable.

Other enginering plastics, described as 'semi-aromatic', are made up of a mixture of straight chains and ring systems, and in consequence are intermediate in chemical, thermal and burning performance. Polycarbonate and the thermoplastic polyesters PET and PBT are the most prominent of these semi-aromatics.

Transparent polymers tend to be grouped together, although there is quite

Fig. 4.3 Automotive petrol filter in nylon

a range of performance and price across the spectrum from polystyrene (a bulk polymer) through SAN and PMMA to polycarbonate, a semi-aromatic which rates as an engineering plastic because of its balance of mechanical properties.

There have been analogous (but less extensive) developments in thermosetting materials. Again, introducing ring systems into the polymer chains makes the material more stable, particularly with respect to temperature. As with thermoplastics, there is something of a 'trade-off'; stable molecules imply not only high performance, but also difficult processing. Polyimide (PI) is an extreme case; of outstanding performance, it is not amenable to standard processing methods.

Table 4.1 is an attempt to draw up a very basic map of the plastics field, simplifying it to ten groupings. Just two essential criteria are considered: the cost, and the upper temperature limit for continuous service (a vital consideration in most engineering applications).

Table 4.1 Polymer groupings: a comparison

Group	Examples	Cost (rel. to PP=1)	Continuous service temp, °C
Bulk thermoplastics	PP, LDPE, HDPE	1–1.5	60–100
Bulk thermosets	PF, UF	1–2	100–160
Transitional thermoplastics	ABS, PMMA, SAN	1.5–3	90–110
Engineering thermoplastics	Nylon 6, 66 & 12; acetal PC, reinforced PP, PPO	2–4	90–130
Engineering thermosets	Polyester, polyurethane, SMC, BMC, vinyl ester	3–5	110–220
Semi-aromatic thermoplastics, etc	PET, PBT, nylon 46, polyarylates	4–8	120–180
Aromatic thermoplastics	Polysulphone, PES, PPS	9–14	150–220
Fluoropolymers	PTFE, ETFCE, FEP	5–20	150–260
Heterocyclics (tp & ts)	Polyimides	15–70	170–260
Aromatic polyketones	PEEK	50–80	250–270

It is evident that enhanced continuous service temperature does not come cheaply. These expensive aromatics, heterocyclics and fluoropolymers nevertheless justify their selection in a great many specialised applications, especially where extreme chemical resistance and fire resistance are needed as well as high temperature performance.

Crystalline and Amorphous Thermoplastics

Crystallinity is important in thermoplastics even when there has been no deliberate orientation, as in fibres and hinges. Crystallinity occurs intermittently in a 'straight' chain polymer wherever the chains can become closely aligned. When a thermoplastic has bulky side chains attached to its carbon 'backbone', close alignment becomes impossible, even with very slow cooling. There is no crystallinity, and the polymers are described as amorphous.

Different property profiles are associated with amorphous and semi-crystalline polymers. Table 4.2 gives some idea of the kind of applications which are appropriate to each category.

Table 4.2 Examples of uses of crystalline and amorphous plastics

Property	Applications	
	Crystalline	Amorphous
Solvent resistance	Nylon for oil filler caps, petrol filters and tubing: PP for fluid reservoirs	PC only suitable for car bumpers when blended with PBT (crystalline)
Fatigue resistance	Nylon for bonnet catches; acetal for door striker plates, PP in kitchen equipt.	
Easy flowing melt	Nylon for cable ties; PP for one-piece car roof and pillar liners	
Transparency	Nylon translucent in thin sections, used for light diffusers	PMMA for rear light lenses; PC for headlamps
Dimensional stability	High thermal expansion	PC/ABS blends for cowl panels; wing panels in PPO/PA
Properties stable with temperature		PES for very high temp. electrical connectors

Plastics and Elastomers

This categorisation cuts right across the others. The term 'elastomer' relates to rubber-like polymers which are readily deformed by low stresses, and revert rapidly to something like their initial dimensions when the stress is removed: the elastic band is an obvious example. Such materials are almost completely elastic in normal use and do not show significant plastic deformation.

Copolymers and Blends

Molecular 'tailoring' can be seen at its most spectacular in these materials. Many thermoplastics are derived from two or more monomers, either as blends or copolymers. For our purpose it will suffice to consider the two 'simple' extremes.

A typical copolymer example is one where a second but compatible monomer has been introduced during the polymerisation. Its effect may be to reduce the crystallinity and so make the melt processing characteristics less critical. In performance terms the effect may be to reduce the modulus and ultimate strength, but to increase the toughness and fatigue resistance. Such copolymers are likely to be intermediate between the two 'pure' polymers, in respect of thermal and chemical behaviour.

The most effective form of a blend is when one polymer is dispersed within another. This can produce big changes in the mechanical properties of the major ingredient, the continuous phase, without affecting its chemical and surface characteristics or its melting point. The two components may be very different polymer types; this helps to ensure that one remains dispersed in the other.

Some of the most important copolymers and blends are these:

Fig. 4.6 Side view mirror on Kaessbohrer Setra bus, moulded in ABS copolymer and ASA-PC blend (courtesy BASF)

Fig. 4.7 Wings (fenders) of the GM Saturn concept car were moulded in PPO/nylon 66 blend. Subsequently PPO/PA wings have been used in Nissan and Renault production models (courtesy GE Plastics).

- Polyolefin rubbers blended with polypropylene to give tough grades for car bumpers, etc.
- Polystyrene added to polyphenylene oxide to give materials (PPO) with much improved processing at lower cost, for business machine housings, etc.
- PBT thermoplastic polyester added to polycarbonate to enhance its solvent resistance, making it suitable for car bumpers.
- Blending of ABS with polycarbonate to give an optimum balance of impact strength, stiffness and heat distortion resistance at an acceptable price.
- Rubber modification of PBT to make bumpers, requiring high impact strength.
- Dispersing PPO in nylon 66, in order to combine the chemical resistance and melting point of nylon with the creep and high temperature rigidity of the amorphous PPO, for body panels which are paintable on-line.

Polymers to Plastics: The Formulation of Compounds

Polymers are very rarely used in their pure state. Commercial plastics are polymers which have been carefully formulated into compounds, with a variety of additives. The manufacture of compounds, usually but not exclusively based on an extrusion process, can be an extremely sophisticated (and expensive) operation. The 'plastics' which the world sees are fabricated not from pure polymers but from these compounds. Overall, some 20% of the material content of plastics is non-polymeric additives (which can have serious implications for recycling). Essentially, there are three kinds of additives.

1. Those designed to make a fundamental change in performance. They are likely to be used in high concentrations; say from 15% up to 60% by weight of the final compound. This category comprises composites and filled compounds. The designation 'composites' in practice refers to polymers reinforced by fibres, predominantly glass. Composites nowadays are a hugely important class of engineering materials.
   When the additives are particulate, the rigidity and form stability are enhanced, and the compound is described as 'filled' rather than reinforced. Filled polymers are important, but compared with fibres, particulates make little contribution to strength. Nevertheless, other properties may be crucially enhanced: for instance, adding copper powder can make phenolic resins highly heat conducting, and powdered lead in polyethylene performs a vital radiation-absorption role in the nuclear industry. Calcium carbonate and other minerals are used in sheet moulding compound and in many grades of glass fibre reinforced thermoplastics.
2. Many additives perform an 'add-on' function, without changing the basic mechanical performance. These include anti-oxidants, to reduce the effects of heat or UV ageing; lubricants or slip additives to improve the frictional or abrasive performance; conducting powders or fibres to impart static dissipation or a degree of conductivity, and of course a great variety of pigments and dyes.
3. Other additives are used (usually in very small amounts) to initiate or control the polymerisation or to modify the conversion process. Their function may have been to control the length of the chains (molecular weight), or to improve the flow rate under high speed injection, or to stimulate crystallisation, or to prevent sticking, say in moulds or in reels of film. The effects of these small additives always need to be carefully assessed: it is possible that a 'performance' additive can affect processing, and that a 'processing' additive can affect performance. Successful formulations are rarely achieved without 'trade-offs'.

Nowadays considerable efforts are expended in targeting each application with the best available material at the best price. To this end, many formulation variables may be deployed at the same time. Blending, copolymerisation, molecular weight selection, reinforcement, toughening, filling, plus all the 'add-on' materials: these are all factors which may be brought into play.

By way of example, Figs 4.8 and 4.9 and the accompanying Table 4.3 show just how precise the targeting and formulation can be, within a single market area. Polypropylene is the most widely used automotive plastic material; we see here that, in order to meet all the requirements, seven different categories of compound based on polypropylene have evolved. (Furthermore within each category, each manufacturer will have developed its own variants.)

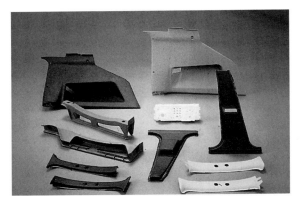

Fig. 4.8 Polypropylene mouldings from Rover car interiors

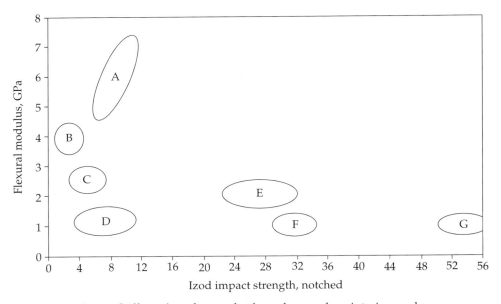

Fig. 4.9 Stiffness/toughness plot for polypropylene interior grades

Table 4.3 Composition of polypropylene interior grades

Class in Fig. 4.9	Formulation	Typical applications
A	Homopolymer + glass fibre	Seat pans; load floors
B	Homopolymer + 40% talc	Heater boxes, instrument backs, parcel tray supports
C	Copolymer + 20% talc	Consoles, door liners, A, C & D pillars, tailgate linings, steering column shrouds, quarter panels
D	Copolymer	B pillars, bootliners, door panels, scuff plates
E	Copolymer, toughened & reinforced	Instrument panels, glove box doors
F	Heavy duty coplymers	Fascia underpanels, boot liners, woodfilled panels
G	Copolymer/elastomer blends	Bumpers and spoilers

5. TRANSFORMATION
The Processing of Plastics

THE GENERAL PICTURE

Shapes can be cut directly from solid plastics, just as they can from wood or metals. Prototypes can be machined from rod stock; washers can be stamped from sheet. The tools are essentially the same, allowing for different thermal and dimensional behaviour. However this is not using the materials to best advantage. The essence of plastics is their plasticity, the ease with which they can be softened and shaped.

'Processing' (or 'converting') covers all the methods by which plastics raw materials are transformed into useful objects. Each involves a great deal of specialist expertise, beyond the scope of this book. Nevertheless it is impossible to understand plastics without appreciating the basic facts of these transformations. No plastic material is ever selected simply because of its excellent properties. There must also be feasibility, which in truth means practicable processing plus acceptable costs.

The choice of process is determined partly by the needs of the final product, viz.,

- Is it required in continuous lengths or in repeated units?
- Does it demand precise dimensional control, and on all surfaces?
- Do the numbers required justify large up-front expenditure?
- Are surface quality and appearance important?

Material properties also affect the choice of process, viz.,

- Is it thermoplastic or thermosetting?

If a thermoplastic,

- Does it soften over a wide or narrow temperature range?
- Does it have a very high softening point?
- Is the melt free-flowing, or is it highly viscous?
- Does it melt at all, or can it only be formed by sintering?

OVERVIEW OF THE PROCESSING METHODS

Table 5.1 summarises the plastics processes. They are classified according to whether they are continuous or discontinuous, and to whether they need high pressure machinery.

Some of these processes have their origins in antiquity; others are very recent arrivals. Some processes are broadly suitable for both thermoplastics and thermosets: in detail however the needs of the two are very different. The reason is that whereas thermoplastics are chemically stable during the processing cycle, thermosets are actually polymerising ('curing') while they are exposed to heat. Any shaping must therefore be performed before curing has gone too far.

There is some overlap between different methods, and the specialist processes are continually borrowing from each other. Nevertheless the basic concepts in plastics conversion are quite distinct.

Table 5.1 Processes for plastics

Raw material form	Continuous process	Discontinuous process	
		High pressure	Low pressure
Monomers and liquid resin	–	–	Casting Reaction injection moulding Resin transfer moulding GRP contact moulding
Powder and paste	Calendering	Compression moulding	Dip coating Slush moulding Fluidised bed coating Rotational moulding Sintering
Granules	Extrusion	Injection moulding Blow moulding	–
Sheet	–	Shaping of SMC & GMT	Thermoforming

There follows a review of the more important of these processing methods, highlighting the potential benefits and the inevitable constraints.

EXTRUSION

Extrusion is the definitive continuous process, where the products are effectively infinite lengths of uniform cross-section. Its origins are to be found in the Italian pasta industry, although the modern extruder is unrecognisably different. Thermoplastic granules are softened and melted in a heated cylinder, and conveyed forward by means of a screw, at a constant speed. The melt emerges through a die cut according to the shape of section required. The final dimensions of the continuous section depend on how much the melt is stretched before it solidifies. Modern extruders are extremely sophisticated, with attention focussing on screw design (including twin-screw constructions for improved mixing and consistency), on design of the crosshead which leads the melt to the die, and on systems for ensuring uniform output.

Some very familiar products are made by extrusion processes:

Cable Sheathing

The melt is drawn on to the cable as it passes at high speed through the die head. Low viscosity grades are used because the melt is supported on the cable core. Telecommunications, automotive cable and domestic wiring circuits all use PVC insulation. A landmark in the development of LDPE was submarine cable, demanding high electrical performance. More expensive polymers are used for cables requiring high resistance to fire or radiation.

Filaments

These are formed by extrusion through a multi-hole die. There are old-established uses like nylon monofilaments for brushes and racquet stringing, and new ones like acrylic filaments for fibre optics.

Tubing and Pipe

These are the products of extrusion through an annular die. Examples are LDPE pipe for water services; plasticised PVC for garden hose and rigid PVC for gas pipes, electrical conduits and rainwater goods; automotive petrol and pressurised air lines in nylon 12, and medical uses (including catheters and implants) in high performance materials.

Extruded Sections

Accurate sections are best achieved in amorphous plastics with a viscous melt: hence the prolific use of PVC in window frames and curtain rails, etc. Housings for lighting installations are extruded in transparent acrylic and polycarbonate.

Flat Film

Flat film is produced by extruding through a slit die and cooling the melt on a chill roll assembly. The main use is of polystyrene and polypropylene for thin-walled food packaging. Biaxial stretching greatly increases the strength and clarity of film; PET and polypropylene are the most successful examples. Above 1 mm thickness or so extrudates of most plastics are rigid sheets. They are used in an impressive array of decorative, building and horticultural products.

Fig. 5.1 Extrusion of acrylic lighting diffuser: the melt emerges as a corrugated pipe, and is then slit and shaped (courtesy ICI)

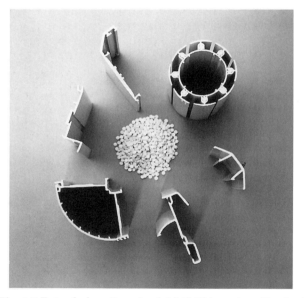

Fig. 5.2 Extruded sections in rigid PVC (courtesy Doeflex)

Tubular Film

This is formed by upward extrusion through a ring die, the thin tube of melt being stabilised and stretched by a positive air pressure (the 'bubble'). The film is drawn upwards and nipped into the 'layflat' form and rolled up for further processing. Blown film is more widely used than flat film, as it can either be slit and used flat, or used as a basis for bag making. High speed production of bags and sacks, especially with low and high density polyethylene, is now a major world-wide industry involving very specialised equipment.

INJECTION MOULDING

Injection moulding is perhaps the ultimate mass production process: it is certainly the best way of producing precision components in very large numbers. Because of the pressures involved, moulding machines are massively constructed, and their capability can extend to completely automatic operation, with the elimination of all finishing operations.

Injection moulding combines the two stages of melting and shaping. Essentially a method for thermoplastics, it involves melting the granulated material in a heated barrel with a reciprocating screw. At the appropriate point of the cycle, the screw also functions as a plunger, injecting the melt under high pressure into a closed mould. Successive mouldings are produced on a concurrent sequence of melting, injection, cooling and ejection, with very precise control of the time and temperature variables.

Both the machines and the moulding tools demand a great deal of quality engineering. Therefore, before the many benefits can be realised, there is a considerable 'up-front' cost. This is the central economic fact of injection moulding. Additionally, much expertise is required to ensure its success, starting at the design stage. Calculations are necessary to relate factors such as component life, numbers required, tool life, rate of call-off, etc, to the number of tools necessary, or (for small components) the number of mould cavities per tool. Small multi-cavity mouldings like those in Fig. 5.3 present different design problems from those of a large intricate moulding like that of Fig. 5.4.

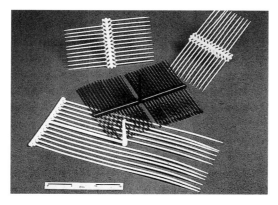

Fig. 5.3 Nylon cable ties moulded in multi-cavity tools

Fig. 5.4 Complex moulding from multi-gated tool: instrument panel for Nissan in polypropylene

Modern injection moulding is very sophisticated. An appreciation of the pattern of flow and its implications for orientation and shrinkage is important, especially when fibres are present. Balancing the flow between different parts of the tool and ensuring that weld lines do not occur in impact-sensitive areas are critical aspects. Computer aided design with flow simulation programs plays a vital role. Computer aided manufacture completes the picture, based on precise control of all the process variables such as injection speed, cycle times and temperatures of melt and tool.

There must be significant benefits to justify all this trouble and expense. If the numbers are large enough, there are benefits indeed:

- No finishing operations.
- Design freedom, allowing new shapes and forms.
- Component consolidation, meaning fewer parts and processes.
- Very high output rates.
- Lower piece part costs.

With conventional injection moulding it is impossible to produce closed or hollow shapes, or ones with excessive undercuts. Very complex shapes can be made by incorporating sliding cores, which in effect change the shape of the tool at certain stages of the process, but these will entail higher costs and longer moulding cycles. There is often a 'trade off' between higher initial costs and higher running costs.

Recent years have seen several ingenious new developments, which have enabled the injection moulding process to break through some of its traditional boundaries:

Structural Foam Moulding

Incorporating a chemical blowing agent in the granules results in a rigid lightweight foam structure. Suitable for large thick section mouldings, this gives a high stiffness to weight ratio. Limitations of the process are a poor surface finish and a longer cycle time.

Multi-Shot Moulding

This technique was devised for moulding automotive rear light lens clusters in PMMA. It is based on a rotary mould assembly, into which the different colours are injected sequentially from separate cylinders.

Sandwich Moulding

Two materials are injected sequentially through a single feeder channel (sprue). The first constitutes the skin of the moulding and the second becomes the core. The original ICI process focussed on a foamed core to maximise the stiffness to weight ratio. A more recent innovation is the use of recycled material for the core, whilst retaining first quality material as the outer layer.

Gas Assisted Moulding

This system achieves a high stiffness to weight ratio without sacrificing surface finish. It originally focussed simply on the concept of the gas being localised in the middle of the thick sections, instead of being dispersed throughout as a foam. This provided big new design opportunities, sometimes with the central cavity having a secondary function (see Fig. 5.5). It is now recognised that the gas channels provide a means of using pressure much more efficiently. As a result the system has become extremely versatile, and is even applicable to thin sections.

Injection Moulding Thermosets

The process is the same in principle as for thermoplastics. The difference is the vital need for accurate control of time and temperature, to avoid premature curing. An important sector is the processing of bulk moulding compound, BMC (dough moulding compound, DMC, in the UK). These materials are usually unsaturated polyesters filled with talc as well as glass fibre of medium length (up to 12 mm).

Fusible Core Moulding

This is a method of producing complex hollow objects by injection moulding. It is basically a 'lost wax' type of process, in which the component is moulded

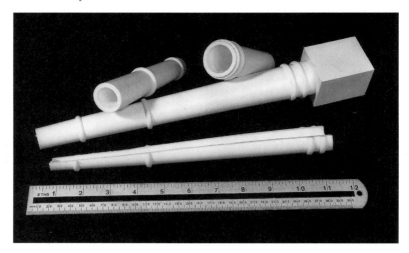

Fig. 5.5 Gas assisted moulding of a chair leg in polypropylene, showing a concentric hole cored out by gas alone, without mechanical cores (courtesy Gas Injection Ltd)

around a low-melting metal core located within a conventional tool. After moulding, the core is melted by induction heating and then drained away, ready to be recast into new cores (see Fig. 5.6). The fusible alloys are expensive and the cores themselves can be extremely heavy: a great deal of production engineering has been devoted to perfecting this method. The first successful application was tennis racquet frames, but the most important now is automotive intake manifolds. These have now been established primarily in glass reinforced nylon 66.

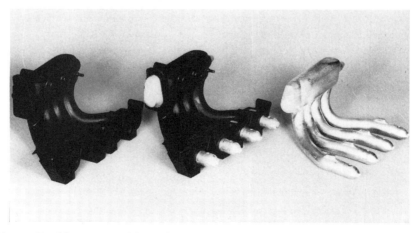

Fig. 5.6 Fusible core moulding of an intake manifold in glass reinforced nylon, in three stages (courtesy BASF)

BLOW MOULDING

Extrusion Blow Moulding

Blow moulding is the simplest way of making hollow plastic articles. It requires materials having a fairly broad melting range and a highly viscous melt. The original method is similar to that used for production of glass bottles, and takes place in three stages. Melt is extruded continuously into an accumulator; tubular preforms (parisons) are repeatedly ejected from this through an annular die into the open blowing tool, whereupon the tool is closed and the preform rapidly inflated.

The process is cheaper than injection moulding because of the lower pressures involved. Furthermore, 'twin shell' forms can be blown as a single entity and then parted, and external screw threads can be moulded in very simply. The disadvantages are that dimensional accuracy is only possible on the outer wall, and that the wall itself may have to be thicker than ideal to accomodate the inevitable thinning at corners. Nowadays however it is possible to blow highly assymetrical shapes without unacceptable variation in wall thickness, by means of computerised parison control. Also, multi-layer blow moulding now makes it possible to combine the benefits of two dissimilar materials. Both of these effects are put to good use in the automotive fuel tank (Fig. 2.9), which is currently the most important engineering application of extrusion blow moulding.

Injection Blow Moulding

This process was at the centre of a remarkable success story: the conversion of PET into bottles for carbonated drinks. Exploiting the unique benefits of a polymer whilst avoiding its deficiencies has deservedly led to a large and profitable industry. There are two quite separate stages: the first is the injection moulding and very rapid cooling of a small preform, and the second is the stretching and blowing of the softened preform. The rapid cooling prevents the PET from crystallising and becoming opaque. Compared with extrusion blow moulding; there is no waste, and the walls are thinner and more uniform.

ROTATIONAL MOULDING

This is the most practicable way of making large hollow containers. It is suitable for small production volumes, because the low pressures entailed can be accomodated with cheap tooling.

In essence the process is very simple. Powdered polymer is placed in a heated mould, which is then rotated along two axes so that the powder is flung centrifugally against the heated wall of the mould, where it fuses, before being cooled. Good control of temperature and of rotation speeds results in a coherent stress-free moulding, with an almost uniform wall thickness and no waste. Rotational moulding is also effective for small mouldings, when the numbers required do not justify the up-front cost of blow moulding equipment. The limitations are that the process is rather slow, and that the necessary fine powders are more expensive than granules. Furthermore, coherent mouldings are best achieved with polymer grades of low molecular weight, which tend to be less tough than the high viscosity grades of blow moulding.

Rotationally moulded dustbins have displaced galvanised steel ones from the domestic market within a very few years. The ubiquitous traffic cone is another very visible example of rotational moulding. However there are well-engineered applications to be seen in more demanding market sectors, such as fuel tanks for heavy lorries and military vehicles, and a variety of corrosion-resistant industrial containers.

COMPRESSION MOULDING

Throughout most of the life of the plastics industry, the basic technique for producing close tolerance parts in thermosets has been compression moulding. Arguably, the industry attained engineering respectability on the back of a host of compression mouldings in filled PF (Bakelite, etc) for electrical applications in the domestic, automotive and industrial markets (Fig. 5.7). The procedure is to fill a two-part tool with powder and apply heat and pressure until the polymer is cured. Compared with injection moulding, the process suffers from longer cycles and additional finishing operations, particularly flash removal.

Fig. 5.7 Automotive mouldings in filled PF, from the 1930s

REACTION INJECTION MOULDING (RIM)

The process involves mixing two low viscosity monomers and injecting them rapidly into a mould, where they polymerise. Thus the shaping of the final product is achieved in a single stage, without intermediates such as powder, granules or sheet. Add to this the fact that complex shapes can be made in low pressure tooling with low energy consumption, and it is easy to explain the success of polyurethanes since they arrived in the 1960s. However, for high volume production, RIM does not compare with injection moulding. Long cycle times are inevitable, because of the time taken to polymerise and release the moulding from the mould. However, the gap is closing as RIM technology improves.

The reaction between the monomers can be modified to produce a gas; in fact some 90% of the world polyurethane market is in foams. There are three distinct types of these, supplying very different applications:

- Low density flexible foams. These are open cell and make excellent cushioning materials.
- Low density rigid foams. These are highly cross-linked polymers with a closed cell structure. Thermal insulation is excellent, so refrigeration is the main application.
- High density flexible foams. These have an 'integral skin' which is virtually solid, and which accurately replicates the mould curface. They are ideal for automotive trim like arm rests and steering wheels.

Polyurethane polymerisation can also yield solid elastomers. These are thermoplastic and are supplied as granules, for use in conventional injection moulding and extrusion. They are excellent engineering materials, with great resilience and extraordinary abrasion resistance. Applications range from bushes, grommets, hammer heads, castors and mechanical rollers, to hoses, shoe heels and ski boots.

Fig. 5.8 Steering wheel with integral skin polyurethane foam (courtesy ICI)

THERMOFORMING

The raw material here is a thermoplastic sheet: it is heated and formed by the application of pressure or vacuum. Only low pressures are involved, so the basic equipment for a large shaping like an acrylic bath is relatively cheap and simple. For the same reason, vacuum forming can be a useful prototyping route for parts which will be ultimately injection moulded. On the other hand, high speed production processes in thin-walled packaging do of course need very sophisticated machinery.

PROCESSING OF REINFORCED PLASTICS

When the reinforcement is performed by short fibres and the raw material is in the form of granules, then processes such as injection moulding and extrusion work very well. Allowance must be made for the reduced flow of the melt and its more rapid solidification, and for possible wear on the moving parts of the machinery. Nevertheless the processes are essentially the same. The injection moulding process, duly modified, also works for short fibre reinforced thermosets such as bulk moulding compounds (BMC).

However, most of the glass reinforced plastics industry is concerned not with the processing of premixed granules, but with the forming of a resin-impregnated chopped strand mat during the actual curing of the thermosetting resin. Whether the method is the original labour-intensive hand lay-up technique or one of its semi-industrialised improvements, the fact is that the methods are very specialised and quite different from anything encountered with unreinforced plastics. These methods will therefore be described alongside the composites themselves, in the next chapter.

6. PUSHING POLYMERS TO THEIR LIMITS
Composites

THE BACKGROUND

This topic is bedevilled with misleading nomenclature. In popular parlance, all technological achievements are attributed to 'fibre glass' or 'carbon fibre', or occasionally 'composites'. Never to 'reinforced plastics', which is what they are. It could all be part of the conspiracy to deny plastics any claim to respectability...

'Composite' can mean anything made up of more than one part. In the world of plastics it means a combination of physically distinct materials where the continuous phase is a polymer. In all the important composites, the discontinuous phase is a reinforcing fibre. Whether natural or synthetic, most of our World's more effective and sophisticated materials are composites.

Whether we consider the natural world of plants and animals, or the sweep of man-made artifacts from prehistoric bricks to reinforced concrete and aircraft structures, the composite material achieves effects greater than the sum of its parts. It is through reinforcement that polymers can be pushed to the limits of their strength and temperature capability.

Without challenging our assertion that the fibre reinforced composites are the important ones, it is worth recalling that before the modern era of glass and carbon fibres, significant progress was made in other types of composite technology. They are still part of the scene today.

LAMINAR COMPOSITES

Laminar composites are typified by plywood, made up of layers of wood with the grain at right angles in successive layers, interspersed with layers of adhesive. Laminating of phenolic resin with cloth was achieved as early as 1912, for machining into grinding wheels, and for such things as automotive timing gears, replacing rawhide. Asbestos impregnated with PF resin was used for brake linings in the First World War, and the motor industry was still using phenolic laminates for camshaft gears in the 1960s.

In transport and the building industry, surface cladding materials first appearing in the 1930s were usually composed of layers of paper with phenolic or amino resin, formed under heat and pressure into decorative panels. An early 'spectacular' was the development of the Mosquito bomber for the Royal Air Force in the later stages of the Second World War: the air frame was essentially plywood bonded with urea-formaldehyde resin.

Impregnated Fabrics

This category is included for completeness, although the fabric, not the polymer, is the major constituent. The best known early example of a treated textile fabric was Charles Mackintosh's 'mackintosh' of 1823, made from cotton cloth impregnated with rubber solution. Modern variants include such diverse creations as industrial seals, flame-resistant conveyor belts in canvas with PVC, and dress-making materials made of polymer-impregnated non-woven fabrics. Automotive interiors feature many 'luxury' cladding materials; a typical example is expanded PVC backed with a knitted nylon fabric.

Fibre Reinforced Composites: The Days Before Glass

The reinforced plastics industry is based on the post-war development of extremely strong and rigid new fibres, particularly glass. However, before the advent of glass fibres the industry experienced some thirty years of solid but less spectacular development with cellulosic additives, particularly cotton and wood fibres.

The operation of adding fibre reinforcement to polymers began in the 1920s, with the use of absorptive additives like wood flour and cotton waste to absorb the gases evolved in the condensation reaction between phenol and formaldehyde. The result was not only a reduction in porosity but also significant property enhancement.

One of the main engines for the growth of plastics between the wars was the compression moulding of filled and reinforced phenolics. The maturing of both the motor industry and the plastics industry owed much to technology developed by organisations such as Lucas, Bosch and Magneti Marelli in moulding distributors, relays, fuse boxes, magnetos, plug caps, etc, in filled phenolics. In the 1930s, with more powerful moulding presses, larger mouldings appeared in the car interior, in the shape of instrument panels and window surrounds (see Fig. 5.7).

The history of technology is littered with great ideas, which (with the benefit of hindsight) we know to have been false. One of the most spectacular of these was Henry Ford's Soya Bean Car. This was part of a grand design to reduce the dependence of the auto industry on steel. Car bodies were to be made from wood-filled phenolic, to be extracted from soya-bean oil. A prototype was made in 1941, with a tubular steel frame supporting 14 WF-PF panels. A considerable weight saving was achieved, in spite of the panels being a quarter of an inch thick. However, the PF was in fact petroleum based, and the cost predictions for the soya bean exercise were extremely unfavoura-

Fig. 6.1 The Soya Bean Car (courtesy Ford Motor Company)

ble. (Sadly, the economics of large-scale extraction of polymer feedstocks from plants usually do turn out to be unfavourable!) The Soya Bean Car is best remembered now for the photograph of Henry Ford with a woodman's axe, demonstrating its vandal-resistance (Fig. 6.1).

Twenty years later, in very different circumstances, the concept was put into practice in a volume car whose production continued for 25 years: the East German Trabant (Fig. 6.2). The 'Trabbie' was derided in the West for its performance and pollution, but its body panels in pressings of PF (not from soya!) filled with cotton waste were an effective solution to a specification imposed by a command economy. However it was clear even before the first Trabant that the mainstream development for plastic composite car bodies belonged to glass fibre reinforced polyester, GRP.

The Early Years of GRP

The story of modern composites really started with the coming together of glass fibres and unsaturated polyester resins. Unlike phenolics, these resins did not need high temperature in order to polymerise. Mouldings could be produced by hand lay-up in low-cost tools, and a good surface finish obtained by laying down a glass-free 'gelcoat' before applying any glass fibre. Early development was catalysed by military needs during the Second World War, specifically for a strong, microwave-transparent material for radomes. Interest spread rapidly to other applications. Small boat building provided some early 'spectaculars', but the supreme success story was the Chevrolet Corvette

80   Plastics: The Layman's Guide

Fig. 6.2 The Trabant

of 1953, which had a body composed of GRP panels, on a steel space frame. The Corvette has continued at significant production levels (of the order of 50 000 per year) to the present day. Unlike some other 'cult vehicles', it has been an active test bed, steadily modernising its techniques for handling reinforced plastic body panels.

Development involving single very large mouldings proceeded steadily, into items like railway coaches and yacht hulls. Around 1970 HMS *Wilton* appeared, the first Royal Navy minesweeper with a non-metallic hull,

Fig. 6.3 High powered small boat in GRP (courtesy BPF)

containing 65 tonnes of polyester resin and 65 tonnes of woven glass rovings. Nowadays many classes of vessel are made from GRP, in the UK amounting to more than 60% of the market.

Progress has been inhibited over the years by the absence of any high speed integrated production systems. However, right at the end of the century, semi-industrial techniques are now being developed which will facilitate the production of very large GRP mouldings. There are hopes of a resolution of the great anomaly of the fibre reinforced plastics industry: that the production techniques fall far short of the materials' performance.

PERFORMANCE OF COMPOSITES: BENEFITS AND LIMITATIONS

The Controlling Influences

Even the simplest of experiments with glass fibre and polyester resin (with its initiator) shows that the reinforced polymer is something very different from either of its components. The fibres of a rigid material such as glass are very efficient at load bearing but (by the same token) are not formable. The polymeric matrix protects the fibres and diffuses the load, and is of course formable. Because it constitutes the continuous phase, it determines the overall character of the composite (in chemical resistance, for instance), and is the major influence in deciding how the composite is to be fabricated.

Irrespective of the fabrication method, the performance of any composite depends on five factors:

Proportion of fibres
Fibre length
Quality of bonding between the fibres and the matrix
Quality of dispersion of the fibres within the matrix
Fibre orientation

The mathematics of all these aspects of fibrous composites is complex: fortunately the significance of each factor is quite clear in commonsense terms.

The proportion of fibres has an obvious importance. Very small concentrations, where the fibres are too far apart to help each other in load bearing, merely act as impurities, upsetting the integrity of the polymer matrix and actually weakening it. Very high concentrations of fibre are also undesirable, when the fibre area becomes too great for the polymer to wet it effectively, so that the performance of the composite becomes inconsistent.

In the same way, the fibres are ineffective if they are too short. Composites work by transferring shear loads from fibre to fibre through the matrix. The ends of the fibres do not contribute to this transference, so there has to be a significant distance between the ends. The length at which the fibres begin to improve the strength, in a particular polymer matrix, is described as the critical length.

The quality of bonding between fibres and matrix comes into this equation as well; if the bonding is good, then the fibres become effective at shorter length, ie the critical length is low. Bonding and critical lengths can be improved by chemical treatment of the interface. Polypropylene is an outstanding example of a matrix whose bonding with glass can be vastly improved by chemical modification.

The quality of dispersion of the fibres within the matrix is what controls the consistency and reproducibility of the composite. The fibres must be evenly distributed throughout the matrix; they must also be fully 'wetted' by it. Without the bonding which results from this wetting, the shear loads can detach the fibres from the matrix, so that the reinforcement becomes ineffective under stress. The enduring success of glass reinforced nylon since the early 1960s is based on the good bonding between the two components and the good dispersion resulting from the extrusion compounding process.

Fibre orientation is the other crucial feature. A designer can use it to maximise the benefits of a composite: equally, ignoring it can make a design go badly wrong. The benefits of orientation are most striking with continuous fibre reinforcement, if the fibres can be located so that the highest strength is in the preferred direction. The problems are most evident in injection moulding. A fast flowing melt induces orientation: the result is the most under-appreciated problem of designing with composites: anisotropy. Aniso-

Fig. 6.4 Photomicrograph of fracture surface, reinforced nylon

tropy is the condition where mechanical and dimensional properties are directionally dependent. It is always best to assume that some anisotropy is present in a fibrous composite, unless isotropy has been deliberately designed in. This can be achieved either by means of a random glass mat, or by constructing laminates with balanced orientation of continuous fibres, like plywood.

Where Composites Score

In the spectrum of engineering materials, fibre reinforced composites occupy a commanding position, combining some of the best features of metals and plastics. Table 6.1 attempts a summary comparison of metals, plastics and composites, picking out the performance highlights. This is a simplistic, qualitative assessment, indicating just two levels of performance: 'better' and 'worse'. The message of Table 6.1 is that composites do indeed enjoy the best of both worlds.

Table 6.1 Fibre reinforced composites: where they score

Property	Metals	Composites	Plastics
Strength	○	○	●
Creep resistance	○	○	●
Low thermal expansion	○	○	●
High temperature rigidity	○	○	●
Low temperature impact	○	○	●
Crushability	○	●	●
Isotropy	○	●	○
Surface finish	○	●	○
Plastic deformation	●	●	○
Lightness	●	○	○
Fatigue	●	○	○
Lubricity	●	○	○
Corrosion	●	○	○

○ = better, ● = worse

The middle section of Table 6.1 indicates the potential problem areas for fibrous composites. Strength can be very high, but the failure mode tends to be brittle. However, well designed composites, with high concentrations of long fibres well bonded to the matrix, will fail by splintering or delamination, without fragmentation.

Anisotropy can never be ignored. Surface finish can also be a problem with composites, depending on the chemistry of the matrix and the method of processing. The relative lack of plastic deformation means that composites are less 'forgiving', ie less able than unfilled plastics to accommodate local stresses due to dimensional errors or uneven loading.

COMPOSITES: A WIDENING SPECTRUM

Much Variety, and Some Confusion

Nowadays the designer can select from an impressive range of fibre reinforced composites. Glass fibre dominates, but carbon and aramid are increasingly important. There are nearly as many matrix varieties as there are polymers: certainly any new high temperature polymer variant will attract new reinforced compounds. The confusion arises because, in the constant striving for faster and more industrialised fabrication methods, one process borrows from another. The old demarcations between different processes and different species of composite become blurred.

By far the most commonly used fibre is glass: by far the most common methods of manufacture are based on a random mat of chopped strands. The original method of hand lay-up is still the most used; this with its various refinements of spray, airbag and pressure bag are classified as contact moulding.

Table 6.2 Glass fibre reinforced composites

CHOPPED STRANDS		CONTINUOUS	
RANDOM MAT	DISPERSED	UNIDIRECTIONAL	WOVEN MAT
Hand lay-up Spray RTM SMC GMT S-RIM	Injection mldg. R-RIM BMC	Filament wound Pultrusion	Prepregs (for structures)

(GLASS FIBRE branches into CHOPPED STRANDS and CONTINUOUS)

Table 6.2 lists the main species of glass reinforced plastics according to the physical form of the fibres in the final product

Random Mat Composites

Contact moulding methods are the most widely used, with hand lay-up as the most basic. This involves covering the mould with successive layers of resin and glass fibre, and then compacting the composite by hand and allowing it to cure at ambient temperature. The method needs no expensive tooling or special skills, and it is possible to make very large mouldings like boat hulls. However the process is very slow and labour intensive and the quality can be variable.

The spray-up process improves the speed and consistency of the actual laying up, and the compacting can be improved by the vacuum bag and pressure bag techniques. These processes are still slow and labour intensive, and of course only one surface can be directly controlled. The next logical stage in development is matched tool moulding, involving cold curing in a two part mould. This is much closer to a modern industrial process, giving higher production rates, and controlling the finish of both surfaces and the thickness of the section.

For mouldings of manageable size, sheet moulding compound (SMC) is increasingly preferred. SMC is a transportable semi-finshed product in sheet form, with a reasonable shelf life. It is shaped by compression moulding, in a much less labour intensive process, making it suitable for assembly line operations like the motor industry. SMC is a coherent sheet of unsaturated polyester resin, compounded with initiator, reinforcing fibre and mineral filler. It has been widely used in the USA for a wide variety of car body panels (mostly 'add-ons'): the world's first all-plastic truck cab appeared in the UK in the early 1970s, with 17 separate SMC panels bolted to a steel frame.

Today semi-structural parts in Europe frequently feature SMC: the front panel of the 1993 Citroen Xantia is a good example of an intricate load bearing component.

There are variations using vinyl ester (for higher temperature capability) or added thermoplastics (for improved impact strength). SMC formulations can

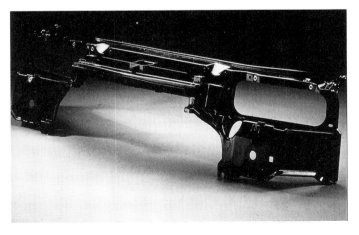

Fig. 6.5 Citroen Xantia part in SMC (courtesy Owens Corning)

Fig. 6.6 Renault Espace

be designed for specific applications with a high directional stiffness, like car bumper beams, by using continuous unidirectional fibres as additional reinforcement.

Resin transfer moulding (RTM) is another closed-mould variant, which has been closely identified with the French motor industry since its introduction as body panels for specialist Matra vehicles. RTM has been used for large area mouldings like bus shelters and coach body panels, and (using epoxy resin) for the side panels of the Renault Espace.

An improved vacuum assisted version is used for the upper and lower halves of Lotus car bodies, which are subsequently joined by adhesive.

Since the early 1980s, glass mat thermoplastics (GMT) have been established as an important alternative. Like SMC, the preform is a sheet composed largely of random glass fibre mat impregnated with resin. Unlike SMC, the resin is a thermoplastic, fully polymerised, and the composite is formed by pressing preheated sheets in two-part cold moulds. There are two different types of GMT: one uses melt extrusion to impregnate a preformed glass mat, while the other involves a paper-making type of process, starting with a slurry of powdered polymer and chopped glass fibres which is then pressed and consolidated. The most frequently used matrix is polypropylene: the process is becoming increasingly automated and industrialised. GMT is extensively used by the motor industry, particularly in 'unseen' unpainted parts such as engine under-panels, radiator support panels and seat backs.

A development with considerable potential is structural reaction injection moulding (S-RIM). This involves placing a glass mat preform in the mould (sometimes with additional unidirectional fibres) and then injecting the monomeric liquids under pressure. Because of the high penetration of the

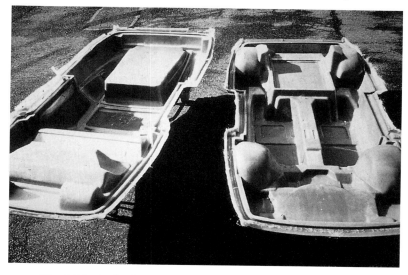

Fig. 6.7 Lotus body construction (courtesy Lotus Engineering)

monomers, the polyurethane as it is formed is very intimately dispersed within and around the glass mat.

These closed mould systems with SMC, RTM, GMT and S-RIM are all more efficient than the older open mould methods, although of course they command much higher initial costs, because of the necessary tooling. They are also more environmentally acceptable, because styrene (used as monomer

Fig. 6.8 Audi seat back in GMT-PP (courtesy BASF)

solvent in the polyester systems) is either much less evident or absent altogether. The refinement of using unidirectional fibres as additional local reinforcement is available for all these methods: however it is generally easier to control the location of the extra reinforcement in the separate insertion processes (RTM and S-RIM) than in the sheet forming processes (SMC and GMT).

Dispersed Fibre Composites

Short fibre reinforced compounds were developed in the early 1960s, with the aim of marrying the properties of hand lay-up composites with the mass production capability of the injection moulding process. This naturally turned the focus on to thermoplastics, especially nylon, which was already established as an engineering material. Initially compounds were made by extruding nylon over a core of continuous glass roving: the properties proved to be inconsistent, however, because the nylon matrix was not wetting the glass effectively. This could only be achieved by direct compounding of glass and nylon in the extruder, with considerable energy input. The result (apart from some badly worn extruders) was a product with fibres which, although short, were extremely well dispersed and wetted.

Early successes for glass reinforced nylon included thin-section coil formers for telecommunications, electrical tool housings (Fig. 1.8) and end tanks for cooling radiators (see Fig. 8.4). This is a good example of how a really effective reinforcement system can stretch the limits of performance of a polymer. Nylon 66 is not inherently a hydrolysis resistant polymer, but nevertheless this application has performed very effectively in a great variety of designs for over 25 years.

Short fibre reinforced thermoplastics, suitable for injection moulding, now constitute a vastly important group of engineering materials, with nylon still the market leader. The application areas are essentially different from the mainstream GRP composites. Injection moulding is most effective for components which although complex are not of very large area, and are required in very large numbers. Glass reinforcement was really a giant step for engineering thermoplastics, because it provided them with strength, rigidity and creep resistance of an order hitherto only associated with metals. The most successful matrices for the engineering sector are those with additional virtues such as outstanding chemical resistance or temperature resistance. The best examples are nylon 66, 6 and 46, acetal, PET, PES, PEEK, PPS and other 'aromatic' polymers. (It is evident from Table 4.1 that every increase in service temperature commands quite a high 'price tag'.)

In the 1990s similar glass-nylon compounds have created a vast new market in intake manifolds (see Fig. 2.12). Glass reinforced polypropylene replaces sheet steel assemblies for washing machine outer drums (see Fig. 3.1) and (as glass mat thermoplastic) automotive front-end structures (see Fig.

Fig. 6.9 Transmission connector in nylon 46 (courtesy DSM Polymers)

2.10). For components like door handles and windscreen wipers, demanding greater dimensional stability and UV resistance, glass reinforced acetals and thermoplastic polyesters (PBT and PET) may be preferred.

Higher performance thermoplastics with glass reinforcement are used in small but crucial electrical and electronic components, where the need is for temperature resistance, precise dimensions and low flammability, often in an aggressive chemical environment. The Volkswagen transmission connector in Fig. 6.9 has to withstand continuous immersion in transmission oil at 160 °C, peaking to 180 °C.

A new development in the 1980s was 'long fibre' variations of dispersed fibre composites. Again with a focus on nylon, these compounds are produced by pultrusion, ie by impregnating continuous rovings with resin and then chopping the resultant lace. When injection moulded the glass fibres are much longer than with the conventional extrusion compounded material (typically 10 to 20 mm compared with 0.5 to 3 mm). The mark-up in strength and rigidity enables these long fibre grades to compete with machined steel in demanding applications. Fig. 6.10 shows schematically how strength and toughness are very sensitive to fibre length.

Although developments in the dispersed fibres market sector have been dominated by thermoplastics since the 1960s, there has been significant progress in thermosets, riding on their generally superior dimensional stability and resistance to burning. A new generation of glass fibre reinforced phenolics has emerged, suitable for precision engineering components in the underbonnet area. Bulk moulding compounds (polyester based mixtures with chopped glass strands and fillers, of composition similar to SMC), are also injection mouldable. They are used for close tolerance electrical components like edge connectors in telecommunications, and in the motor industry for small body components and headlamp housings.

Dispersed glass fibres are also used to enhance the strength and rigidity of polyurethanes in the RIM process. The improvement is necessarily rather

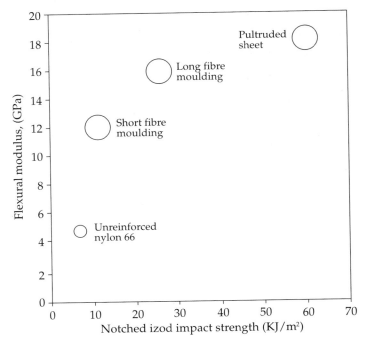

Fig. 6.10 Effect of fibre length in glass reinforced nylon 66

limited because the process of injecting liquid monomers can only tolerate very short glass fibres.

CONTINUOUS FIBRE COMPOSITES: UNIDIRECTIONAL

The technical 'cutting edge' must lie with composites which exploit the outstanding strength of continuous fibres whilst ensuring that they are fully wetted by the matrix. The superlative strength of fibres (in the direction of their length) means that there is an obvious advantage in preserving the length as much as possible. Pultrusion is employed with thermoset resins in applications such as industrial pipework and automotive leaf springs. Continuous fibre prepregs have been successfully applied to superior (and high added value) sports goods like fishing rods and golf clubs. Pultruded reinforced polyester has in recent years become established as a structural material in civil engineering.

Recently, thermoplastic pultruded prepregs composed of polypropylene and continuous glass fibre have been developed, and are likely to constitute a new class of relatively low-cost structural composites.

Filament winding is a process of great potential, which so far has been largely confined to basically cylindrical articles (Fig. 6.11). In this method continuous tows of fibre are passed through a resin bath and then wound on

Fig. 6.11 Filament winding

to a mandrel, which is subsequently removed. Automotive drive shafts have been targeted for many years, for the perceived benefits of reduced noise and vibration, as well as weight reduction. The four-wheel drive version of the Renault Espace went into production in 1993 with a filament-wound shaft, using an epoxy matrix and fibres of both glass and carbon.

Woven Fibre Composites

Neat, exclusive classifications are not always possible in the composites field. All the systems which normally use pre-set random mat are capable of being adapted to woven mat, which tends to be used in high performance, high cost applications.

The aircraft industry is prepared to pay handsomely for weight reduction, and consequently there has been considerable interest in high performance structural composites, based on glass and carbon fibre prepregs. In spite of a recession and reduced defence spending in the 1990s, this market continues to grow. The polymer matrices are thermosets with high temperature resistance and low flammability (usually epoxy), sometimes toughened by the addition of aromatic thermoplastics. Skins of aircraft wings and tailplanes are increasingly being manufactured in this way. In this market complex designs can be executed with great precision. A good example is an aircraft propellor blade with reinforcement comprising a woven glass fabric, with additional oriented strength imparted by unidirectional carbon fibres. A spectacular early example of an all-composite airframe was the Beechcraft Starship of 1988. The basic structure was a carbon fibre epoxy laminate over an aramid honeycomb core.

Formula One racing cars may enjoy even more extreme cost parameters than aircraft. A typical body shell nowadays will be made from a structure similar to the Starship, eg a double skin of carbon fibre reinforced epoxy, with a non-combustible core of meta-aramid fibre. These exotic structures are certainly effective: in recent years there have been many examples on the racing circuits of their life-saving capability.

The technical performance of these structural composites has been repeatedly proven, but the costs are quite unacceptable for volume production.

MARKET SIZE AND GROWTH

Precise figures are difficult to obtain, partly because the terminology differs from one market to another. There is no doubt however that the market continues to grow, and that most of the growth is in thermoplastics. The most spectacular thermoplastic growth is in the automotive industry, in short fibre glass reinforced nylon for intake manifolds, and in polypropylene-impregnated glass mat for front-end structures. However, traditional thermosetting resins are enjoying a surge of new applications in civil engineering, where composites are being used to strengthen existing concrete bridges, and also to produce maintenance-free cladding for new structures like bridges and seawalls. Mass transit vehicles and wind turbine blades also look like providing new high growth applications.

In Western Europe the total market for reinforced plastics is around 1.5 million tonnes, about one third of it in injection moulded thermoplastics. Worldwide, the fibre reinforced composites business is huge. In 1998 it stands at rather more than 5 million tonnes, with a value of around $150 billion. From a standing start in the late 1940s, this is remarkable.

7. MAKING IT HAPPEN: PART ONE
Identifying the Needs and Choosing the Material

TRANSLATING PROPERTIES INTO EFFECTS

Most plastics applications are based on experience; building on what has gone before. Double walled sheet extrusion develops from simple sheet extrusion, success with blowing bottles shaped like glass bottles leads to technology for bottles with built in handles, and so on.

However, when new ground needs to be broken, with an application which may not have existed before, and certainly not in plastics, then life can be much more difficult. How then can we set about getting the desired new effects, armed only with the published material data?

This chapter considers the decisive properties, indicating which of them are truly unchangeable, and which are affected by form and dimensions and by processing conditions. Awareness of this is a necessary prelude to material selection.

Ensuring that the chosen material performs to its full potential is another story. This is very much a function of good design and good processing, and is the subject of chapter 8.

THE PROPERTIES

Fundamental Characteristics

Every plastic material has certain attributes which are essentially constant. Obvious examples are thermal properties like softening and melting points, and chemical properties like permeability, absorption and solvent resistance. Other things however are less constant.

Design Dependent Characteristics

With any material, the strength and rigidity of a component are profoundly affected by form and dimensions, particularly thickness. It may be possible to equal the stiffness of, say, an expensive high modulus material by employing a cheap lower modulus one in a thicker section. However, the choice is likely to be complicated by processing factors. Increasing the thickness may prolong a moulding process or induce internal strains, and reducing it will make any melt flow process more difficult. Much of the art of material selection lies in achieving a balance between the demands of performance and processing.

Many properties, like the electrical insulation parameters, are directly thickness dependent. Thickness is also very relevant to burning performance, with thick specimens showing better resistance than thin ones. Transparency is naturally thickness dependent, the more so with translucent plastics like nylon or polypropylene than with truly transparent ones like polystyrene or polycarbonate.

Surface Based Characteristics

Ageing processes are surface based, in that oxidation initially affects only the surface molecules. However, continuing exposure (whether to UV light, radioactivity or simply air at high temperature) progressively affects molecules below the surface. The embrittled surface layer becomes thicker, and normal expansion and contraction can turn crazing into cracking. Clearly the effects of ageing will be more destructive for a thin section than for a thick one.

There are other surface properties which apparently can be precisely defined, such as coefficient of friction, abrasion resistance and electrical tracking resistance. Such precision is really confined to controlled test conditions, however. In practice behaviour depends on the purity and integrity of the surface.

FUNDAMENTAL CHARACTERISTICS: THE MATERIALS COMPARED

Lightness

Plastics materials are all relatively light weight. Most have a density between 0.9 and 1.2. Important exceptions are PVC at 1.38, some sulphur-containing materials at a similar density, and (because of their extremely compact structure) fluoropolymers at around 2.1. Adding glass fibres or mineral fillers to engineering plastics increases their density to around 1.3 to 1.6.

The advantages of lightness in everyday use are obvious, whether in terms of laying gas pipes or carrying the shopping home. However the dominant factors nowadays are the ecological ones, stressing the energy benefits of lighter moving parts in mechanisms, of lighter vehicle loads and of lighter vehicles. The various plastic foam industries (chapter 2) are flourishing, and their environmental advantages are increasingly apparent. At this level the details of density are less important than such things as impact strength and thermal conductivity.

Upper Temperature Limits: Short Term

Plastics are still sometimes used in applications beyond their proper temperature range. Hence the familiar domestic incidents such as the formaldehyde smells from overheated and under-ventilated PF and UF electrical fittings, the yellowing of badly sited LDPE lampshades and the melting of kitchen utensils left on a hotplate.

Thermosets, with their immobile crosslinked molecules, are fairly stable with respect to temperature. In general, basic characteristics like modulus will deteriorate only slightly until temperatures become high enough to induce chemical changes or charring.

Thermoplastics are much more sensitive to increasing temperature. The most fundamental characteristic defining the high temperature performance of a thermoplastic is its glass transition temperature (Tg). This is the point at which significant internal molecular movement becomes possible. Above the Tg the material will readily creep under load and so lose its effectiveness as an engineering material.

There is an important distinction here between amorphous thermoplastics and semi-crystalline ones. With amorphous materials the Tg is effectively the softening temperature; however with semi-crystalline plastics the change takes place in two stages. Their amorphous regions achieve some mobility at the Tg, but the more compact crystalline parts retain their integrity up to a higher temperature. These semi-crystalline materials (like nylon) have very limited load-bearing capability above their Tg, but things change dramatically when reinforcing fibres are added. Semi-crystalline plastics can then be used at temperatures far above their Tg, and close to the crystalline melting point, Tm. Figure 7.1 compares the effect of temperature on modulus, for amorphous and crystalline plastics, with and without reinforcement.

The Tg is not a practical indicator of the temperature limits for composites, because it is not affected by reinforcement. The best indicator of the short term upper limit is the heat distortion temperature (HDT). This has no fundamental significance, but it is a good practical guide to the upper working limits. HDT comparisons show that the best performers as the temperature rises are the reinforced crystalline thermoplastics.

Upper Temperature Limits: Long Term

The reason for differentiating between short and long term temperature stability is oxidation. When oxidation embrittles the surface, the effect is cumulative: it is irreversible, and furthermore every 10 °C rise in temperature approximately doubles the rate of deterioration. Temperature and time of

96 Plastics: The Layman's Guide

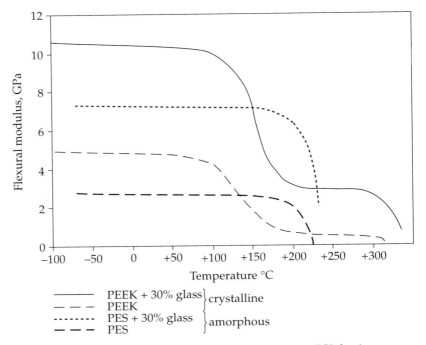

Fig. 7.1 Change of modulus with temperature (ICI data)

exposure are thus intimately connected. Fibre reinforced materials are much better able to withstand oxidation, because fibres restrict crack propagation.

The effects of oxidation are reduced by adding antioxidants (also known, confusingly, as heat stabilisers). They function by preferentially absorbing the oxygen, so their effect is to delay oxidation rather than to prevent it. Their optimum concentration as additives is quite small; too much of them can affect the colour of the material and its chemical and electrical performance.

In defining long term temperature resistance, the concept of 'continuous service temperature' is used, often in a vague sort of way without the key functions being defined. (The Underwriters' Laboratory Index system does permit precise definitions, however.) Great care has to be taken in applying 'continous service temperatures' based on different kinds of components working in different environments. As always, like must be compared with like!

Short term and long term temperature limits are brought together schematically in Fig. 7.2. Two conclusions can be made:

1. Short term stability, as measured by HDT, is always improved by reinforcement.
2. Long term stability, as measured by continuous use temperatures, is much more a function of polymer type. Generalising, the best structure for

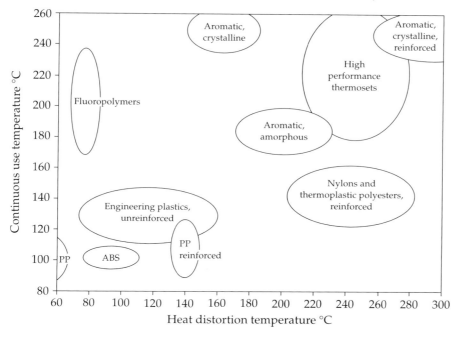

Fig. 7.2 Short term and long term temperature limits

continuous high temperature performance is a molecular chain made up of very durable ring systems. (There are two types of such rings; the 6-membered 'aromatics' and the 5-membered 'heterocyclics').

Heterocyclic polymers containing the 'imide' grouping tend to exhibit the best resistance to thermal oxidation. The inherent advantage of thermosets, in not melting but degrading to a bulky carbonised solid, is spectacularly exploited in space vehicle heat shields. Currently the 'ultimate' heat resisting material is polybenzimidazole (PBI), which keeps a good degree of structural integrity even at 1000 °C.

Once again it is important not to be too reliant on published information. Data are usually quoted for standard thickness (often ⅛ inch, or 3.2 mm); for sections thinner than this, the actual ageing performance is likely to be worse than expected.

Resistance to Ultra Violet Light

The effects of UV light have long been familiar in the fading of dyes and pigments in fabrics and paints. UV light affects polymers by oxidation, in much the same way as heat ageing. The first visible effect is discoloration, in the direction of yellow and then brown: this is followed by microcrazing, leading to a serious loss of resilience in the surface.

Microcrazing can be exacerbated by rain erosion. This causes 'chalking' when white pigments are present, with loss of surface gloss. Reinforced materials tend to lose their good looks sooner, but mechanical deterioration is slower because of their better resistance to crack propagation. UV ageing can be slowed down by added stabilisers, which absorb energy preferentially (as with heat ageing). Carbon black is particularly effective.

Exterior automotive plastic parts often show their age by fading and microcrazing (hence the increasing use of high gloss automotive paint for these parts). Many of the early plastic bumpers in SMC and door-mounted mirrors in nylon in the 1970s faded very rapidly because they were blackened with dyestuffs (which gave a good initial colour) rather than with carbon black, which lasts much longer. Fortunately, that phase, which did nothing for the image of plastics, has long passed into history.

The aromatic materials are again the materials with the best intrinsic resistance to ageing. However, of the plastics in everyday use (at everyday prices), the ones with the best UV resistance are PVC, as in window frames and drainpipes, and polymethyl methacrylate (PMMA), as in automotive rearlights and road signs.

Just as with heat ageing, UV ageing is an effect which progresses inwards from the surface. It is unwise therefore to expect good outdoor performance from a very thin section in any plastic material.

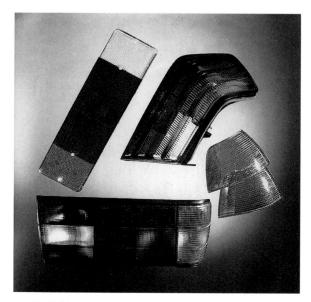

Fig. 7.3 PMMA in automotive rear light lenses (courtesy BASF)

CHEMICAL RESISTANCE

The resistance of a thermoplastic material to chemical attack is one of its most fundamental characteristics. However, a designer looking for precise quantitative data is doomed to be disappointed. Suppliers' literature is necessarily vague and generalised. The problem is that what happens in any specific case depends on many factors: the temperature, the concentration of the active agent, whether the system is at rest or agitated, whether the contact is continuous or intermittent, and whether the material is being stressed. Stress can be particularly damaging to amorphous plastics. The essential differences between the chemical types are summarised in chapter 4.

Figure 7.4 demonstrates the use of glass reinforced polyethersulphons (PES) in dialysis pump components, requiring frequent washing and disinfecting in strong reagents at high temperature. Here the potential chemical weakness of an amorphous polymer is secondary to the good resistance derived from the fully aromatic structure.

RESISTANCE TO BURNING

The flammability of plastics has always been a major concern: with the widespread use of cellulose nitrate in the early days of the photographic industry this was well founded. Nowadays the key end-users like the aircraft industry exert a very tight control on material selection. Sadly, lessons are frequently learned the hard way; this was the case in the early years with nylon fabrics, which burn very easily unless they are surface treated. The 'bottom line' is that all polymers are based on carbon–carbon links, and are therefore inherently combustible.

Most polymers have a higher calorific value than coal or wood. Some have

Fig. 7.4 Dialysis pump components in PES

better flammability resistance than others: all can be improved in some degree by additives; but there is never a simple answer to the question 'How well does a plastic material burn?'

Specifications for flammability can assess many different criteria, viz.:

Ease of ignition
Rate of flame spread
Duration of burning
Oxygen needed to sustain burning
Presence of burning drips
Smoke evolution
Evolution of toxic gases

Individual user industries naturally attach different priorities to these criteria. The aircraft industry places the highest importance on smoke and toxicity: the appliance industry is mainly concerned with the inadvertant ignition of electrical insulators, and the prime concern of the motor industry is the rate of flame spread.

Materials should always be evaluated in a relevant way. Factors like the thickness of the specimen, its aspect in relation to the flame, the type of flame and availability of oxygen are all critical. Obviously, a thin sheet held vertically in a well ventilated area with a flame applied to the bottom edge represents the worst case. A match tossed into a pile of dry wood shavings can lead to a conflagration; but a solid oak door makes an excellent firebreak.

All plastics burn. Nevertheless, when polymers are compared under identical conditions, we can find big differences in behaviour. Thermosets have an inherent advantage over thermoplastics because they do not melt. However with some of them there is a risk of smoke and inflammable volatiles being released at high temperatures.

The guideline for thermoplastics is this: the greater the content of stable ring systems in the molecule, the greater the inherent fire resistance. Hence the fully aromatic polymers like PES, polyphenylene sulphide (PPS) and polyether-etherketone (PEEK) perform well against all the tests. Next in line are the semi-aromatic engineering plastics like PC, and the polyesters PBT and PET. Nylons and polyolefines are rather more flammable, but when properly formulated can meet most of the specifications of the electrical appliance industry.

It is in this electrical appliance sector that the formulation of 'flame retardant' grades of polymers has been most effective. The surge of developments in appliances for the kitchen, in automotive electrics and in electronics in the office are largely dependent on special formulations of engineering plastics. As always, there is a price to pay. Major improvements in burning behaviour can often only be achieved by additions of heavy inorganic oxides, which are liable to cause brittleness.

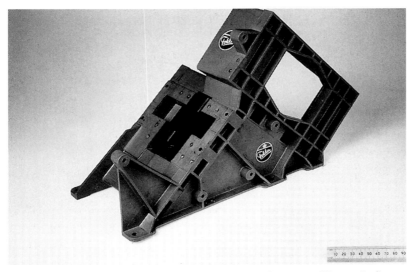

Fig. 7.5 Fokker aircraft component in PEEK (courtesy Victrex Ltd)

Some additives will increase the hazards of toxic gas evolution on burning. Halogens, whether for example as chlorine in PVC polymer or bromine in flame retardants, will produce acid fumes at high temperatures. Nitrogen-containing polymers such as nylons and polyurethane will burn to yield cyanide traces, just as wool does. It is important to be aware of these risks, whilst remembering that the total quantity of active ingredients present may be harmlessly small. The fully aromatic engineering plastics, already noted for their high continuous use temperatures and chemical resistance, can meet many of the flammability specifications without needing extra additives. These materials, when they do burn, exhibit low toxic gas and smoke evolution (which endears them to the aircraft industry). Figure 7.5 shows an elevator control bracket from the Fokker F 100 aircraft moulded in carbon fibre reinforced PEEK.

It should be appreciated that, in any fire situation, the inherent flammability of the materials is only one of many factors. The design of the structure; the location, aspect and thickness of the items of furniture and decor; ventilation, availability of exits and access to emergency services: these are all factors which help to determine whether an incident develops into a catastrophe. And always, whenever a conflagration creates conditions of oxygen deprivation, the biggest hazard is asphyxiation by carbon monoxide.

FRICTION AND WEAR

Test data on coefficient of friction or abrasion resistance do not easily translate into real situations. Neither these parameters nor surface hardness are constant unchanging material characteristics. Hardness depends on the method of surface preparation, and what happens in practice is affected by the presence of trace lubricants (including water), by surface roughness, by the nature of the contacting surface, and by the load exerted. What happens to the dimensions and to the abrasion resistance as the temperature rises are particularly important. In any case, the basic differences between polymers are outweighed by the extra effects achieved by special formulations (see chapter 2).

Transparency

Total glass-like clarity can only be found in amorphous thermoplastics. The best known of these are polystyrene, polymethyl methacrylate (PMMA), polycarbonate and styrene-acrylonitrile (SAN). Polyethylene terephthalate (PET) is normally crystalline, but it can be made amorphous and glass-clear (as befits its use in the carbonated drinks industry).

Polystyrene is the cheapest of the transparent polymers. Optically excellent, it is however somewhat brittle. It was one of the few freely available plastics in the years following the Second World War: it acquired a 'down market' reputation in this period when the industry was distinctly immature in terms of design and production technology. Arguably, much of the bad image of the plastics industry can be attributed to the misuse of PS in those early years. SAN is a modification which increases the toughness without losing transparency, but at a higher price.

PMMA, after a successful wartime career as a material for aircraft canopies, did not have an image problem. This, together with its superior chemical and scratch resistance and toughness, and in spite of a higher price, helped to establish cast acrylic sheet as the material of choice for numerous building and display applications. When granular compounds were developed they soon became favourites for extruded lighting diffusers and a host of injection moulded transparent components for appliances and fittings. PMMA is used for nearly all the world's automotive rear light lenses, bringing with it a substantial new technology in multi-colour moulding (see Fig. 7.3).

'Top of the Range' as a transparent engineering material is polycarbonate. It boasts a somewhat higher temperature limit than acrylic and much higher impact strength. It is widely used as security screens and in impact-sensitive domestic appliances and, since the legal objections were removed in the mid-1990s, in automotive headlamp lenses. It requires coating with silicone, however, in order to match the abrasion resistance of glass (see Fig. 3.3).

Many plastics, not essentially transparent, can be usefully translucent in key applications. The automotive industry uses nylon and PP as diffuse lighting lenses, and for containers like brake fluid reservoirs which need a visible level reading. Some amorphous aromatic polymers like PES are sufficiently transparent and chemical resistant to be put to good use in medical equipment needing repeated sterilisation.

Fig. 7.6 Bobbins in nylon 46 (courtesy DSM Polymers)

ELECTRICAL PROPERTIES

All plastics are good electrical insulators. PTFE, because of its dense, compact structure, is the best of them all. Others like PES are exceptional in that their insulation performance hardly changes with temperature. However, most applications experience only low electrical stress and are satisfied by lower cost plastics. The choice is made on non-electrical considerations such as fatigue resistance, flammability or dimensional stability. Hence nylon is used in automotive circuit connectors and as insulation blocks in automatic railway track signalling because of its resistance to mechanical abuse, rather than its modest electrical perfomance.

Electronic systems make good use of small moulded sensors, connectors and printed circuit board components. These may be specified for immersion in hot oil or air for perhaps 5000 hours at 160 °C or more. Again, materials like PPS, nylon 46, PES and PEEK compete for these vital components on the basis of their chemical and creep resistance rather than their electrical performance.

FORM DEPENDENT PROPERTIES: THE MATERIALS COMPARED

(NB: The fundamentals of the stiffness and strength of polymers, and their response to time, temperature and stress, were examined in chapter 4. This section looks at differences between polymer groups and the changes brought about by additives.)

STIFFNESS AND STRENGTH

The rigidity of a component is highly dependent on design features such as wall thickness, ribbing and box sections. Nevertheless the stiffness and strength variations between different classes of materials are very considerable. The logarithmic plot of Fig. 7.7 confirms this, and also demonstrates that between materials the variations in strength are greater than those in stiffness. Within a group, however, the rigidity differences between individuals may be less important than the design features.

Figures quoted for strength can be even less meaningful than those quoted for stiffness. The visco-elastic nature of polymers, leading to two kinds of failure – ductile and brittle – was described in chapter 4, along with the idea of comparing materials in terms of a 'design stress'. If this incorporates creep data, it is much more valid than using quoted single point data for 'tensile strength', etc. Comparisons between materials must also be made at realistic temperatures.

The importance of temperature is seen in Fig. 7.1. The behaviour with temperature varies very much from one polymer to another, and particularly between crystalline and amorphous polymers. This also shows that fibre reinforcement can make drastic changes in stiffness; particularly so with semi-crystalline polymers. Fig. 7.8 is an example of the 'flat' temperature curve of an amorphous polymer being put to good use. This is a connector in unreinforced PES located in a very hot part of a BMW engine compartment. Unlike semi-crystalline polymers, it is both rigid and resilient under these conditions.

In general, what goes for stiffness is also true for strength. The big difference is that whereas any 'rigid' additive will increase the stiffness, this is

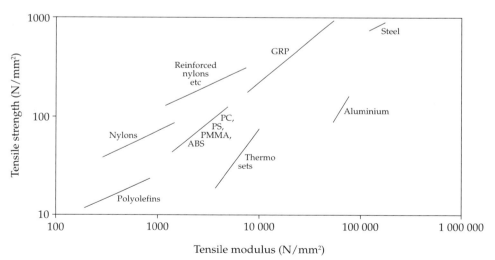

Fig. 7.7 Strength and stiffness of materials

Fig. 7.8 Automotive electrical connector in PES

not the case for strength. Unless the additive is completely wetted and evenly dispersed, it will act as a stress concentration and could lead to failure.

TOUGHNESS

In the language of plastics, toughness means resilience, or the ability to absorb energy; it is characterised by impact strength. The impact strength of an article is a function of many factors, not all related to the material. There are two distinct stages in achieving a component of the appropriate toughness: the first is to use a material of inherently high impact strength: the second is to ensure that the component is free from defects which could cause catastrophic failure. In practical terms, the second stage is more crucial than the first. Unlike stiffness, which can be enhanced in many ways, with toughness the overriding consideration is to avoid the many factors which can undermine it. Avoiding brittle failure, by correct design and processing, is a major feature of the next chapter.

Figure 7.9 reveals how impossible it is to characterise the impact strength of a plastic by a single figure. The effect of temperature on the impact strength of nylon 66, with and without fibre reinforcement, is compared at different degrees of notching.

There are some polymers which fail in a brittle manner under all practical conditions. Others are ductile under all practical conditions: rubber and elastomers are in this category, and are often used as additives to improve the toughness of other polymers. These are the rare extremes, however. Most thermoplastics are at different times either ductile or brittle, depending on the temperature, the rate of loading and the molecular weight of the polymer.

There is one extra problem. Different impact tests compare individual materials very differently. Unless there is relevant application experience to

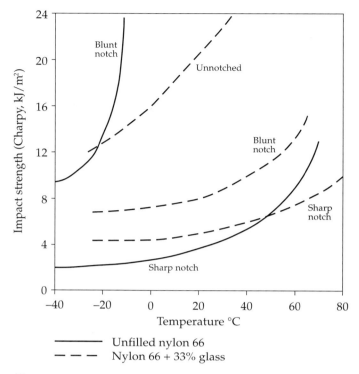

Fig. 7.9 Changes in impact strength with temperature and notching (ICI data)

draw on, the initial material selection tests should always be backed up by tests on finished components.

MATERIAL SELECTION

The Polymer Industry's rapid rise to maturity has been marked by much homespun philosophy, with its occasional failures being excused by 'one-liners' like 'Horses for courses', and 'There are no bad plastics, only bad applications'. Nevertheless there has always been a tendency with a new material to launch it into all and sundry applications. The 'Parkesine' exhibit at the Great International Exhibition of 1862 included several experimental samples, with this confident inscription:

'In the case are shown a few illustrations of the numerous purposes for which it may be supplied, such as Medallions, Salvers, Hollow Ware, Tubes, Buttons, Combs, Knife Handles, Pierced and Fret Work, Inlaid work, Bookbinding, Card Cases, Boxes, Pens, Penholders, etc'.

Most new polymers are subjected to this 'catch-all' process. Examples where the initial applications survive into maturity are quite rare, like the celluloid knife handle, the casein shirt button and the polypropylene hinge. Engineering applications of plastics are nowadays controlled by the end-user industries, and over the years intricate and complex standards and specifications have been meticulously compiled.

Selecting a material for an application tends to be an unstructured and disorganised operation. Experience is a big factor, in particular the designer's love of certain materials and the history of the component. Intuition and prejudice also play a part. Ideally, however, (and quite practicably), we should envisage a sequence like this:

1. Analyse the functions of the component.
2. List the key effects required.
3. Translate these effects into specific properties of materials.
4. Identify the upper and lower limits for each property.
5. Prepare a shortlist, using a database.
6. Make final selection by testing actual components.

The sequence is only valid if each stage is performed before the next is put into operation. This chapter has been focussed on stage 3, the translation of effects into properties. Putting precise values on these properties (stage 4) is not always possible, but much can be achieved with semi-quantitative ratings comparing with known materials.

The preparation of a shortlist (stage 5) is essentially a process of elimination. An experienced 'materials man' would steadily eliminate the unsuitable candidates; those that are too brittle, too flexible, with the wrong chemical performance. Today this selection by elimination is likely to be done with a computerised database.

There are many available databases. They have been devised with different objectives: the strength of some lies in the breadth of choice, while others have strength in depth over a narrower field. Their most vital benefit is to provide the designer with data from different sources and different suppliers which are directly comparable, using identical test methods and identical units. The computer revolutionises the scope, speed and efficiency of the selection process; nevertheless it is meaningless if the data are not comparable.

The database can prepare a meaningful shortlist very quickly. The method is to specify the upper and lower limits for the key properties, and ask the computer to isolate the materials falling within these bounds. However, two important points need to be watched, The first is that some relevant materials may be missing from the database. The second is that the information fed to the computer may have been inadequate. As always, the 'garbage in – garbage out' rule prevails. The more elegant the presentation of the computer output, the harder it is to accept that the input may have been flawed.

Unless both the application and the shortlisted materials are very familiar, the final stage (6) must involve some practical testing of a relevant prototype.

The reason is that some aspect of behaviour will have been omitted from the database. It may be unquantifiable, like 'snap-fit-ability' or environmental stress cracking in an unusual medium; or it may simply have been forgotten. This is more than just a safety net: the component may subsequently be the subject of a cost reduction exercise or a product modification which 'changes the goalposts'. The value of a soundly-based shortlist with a few viable alternatives will then be apparent.

8 MAKING IT HAPPEN: PART TWO
Getting the Best from Each Material

THE PHILOSOPHY

In the words of the old song;

> 'You have to Ac-cen-tuate the Positive-
> E-lim-inate the Negative'

The success of plastics rides on the positives of versatile performance and wide-ranging forming techniques, giving high consistency at low cost. However these all count for nothing against the negatives of poor design and bad processing.

The other important philosophical point is that design and production are basically inseparable. Good design must include a keen awareness of the intended production process, and the choice of process is itself affected by the key design features.

GOOD DESIGN

The principles of good design are universal and relevant in some degree to all kinds of material. However, certain aspects of the behaviour of plastics single them out from other materials, in particular metals.

Design Freedom

- Freedom of shape and form: Injection moulding allows all the design freedom of die casting, but with far fewer finishing processes. Blow moulding can achieve hollow containers in intricate shapes in a single operation.
- Rational design capability: Processes involving matched tools allow the strength and stiffness of components to be controlled by ribbing and by changes in wall thickness.
- Localisation of characteristics: Fracture points can be designed in, eg in single-use trim fasteners. More importantly, most forming processes for composites allow reinforcing mats and unidirectional fibres to be located precisely to maximise strength and stiffness.

- Inclusion of inserts: Design solutions combining the best features of both plastics and metals can be achieved by inserts such as brass screw threads, steel bearings, and steel cores for close-tolerance gears.
- Molecular orientation: Molecular orientation can be induced in semi-crystalline polymers, using the orientation-by-stretching trick which is the basis of the synthetic fibre industry in nylon, polypropylene and PET. The exploitation of this was illustrated in Fig. 2.1.
- Stress relaxation: Compared with metals, plastics are very 'forgiving'. In gear teeth, load spreading transforms point contact into line contact, and line contact into area contact. When experiencing local overload, plastics tend to accomodate and redistribute the stresses, by deformation. At low stresses, this deformation is elastic and recoverable.

Avoiding Dimensional Problems

Failure to allow for dimensional changes probably causes more malfunctions in service than any other factor. There are six main causes of trouble:

1. Thermal expansion
   Any 'mixed' assembly, such as a plastic cover on a metal base, is vulnerable because of differential thermal expansion. Coefficients of thermal expansion are frequently quoted as a single figure. This is inadequate for plastics in general, and for composites it is highly misleading. Because molecular movement increases with temperature, so does thermal expansion.
2. Shrinkage
   Polymers shrink as they cool from the melt, so that the dimensions of the moulding are less than those of the mould. This creates considerable problems in injection moulding, particularly in predicting the precise shrinkage of crystalline plastics, and especially when fibrous reinforcement has caused orientation. Uneven shrinkage results in distortion, so when the general level of shrinkage is higher, so is the distortion. Shrinkage is also affected by the temperature, pressure and time settings during the moulding process. Mould design is a highly skilled operation; shrinkage prediction is perhaps its most complex feature.
3. After-shrinkage
   Rapid cooling from the melt, particularly in thick sections, causes irregular molecular arrangements across the section. The result can be additional shrinkage, accompanied by distortion. The only cure is an annealing operation, by slowly heating to a temperature some 10–20 °C higher than the highest temperature expected in use.
4. Anisotropy
   This is the directional dependence of properties and dimensions, exacerbated by high speed flow (as in injection moulding). It is particularly marked with fibrous composites, where it affects not only thermal expan-

sion and shrinkage, but mechanical and electrical properties as well. Anisotropy is less of a problem with random mat composites like SMC, GMT, and S-RIM.
5. Absorption
Some plastics absorb particular fluids, with consequent dimensional change and deterioration in properties. The only corrective is to avoid unsuitable combinations of polymer and solvent. Rather different is the case of the nylon family, where the polymer will absorb water until it achieves equilibrium with the ambient humidity. This is a precise and predictable process, but failure to anticipate it can lead to seized bearings and failed gear trains.
6. Creep
This is a much more significant in plastics than in metals, and therefore can never be ignored in any long term or high load situation.

Enhancing Rigidity

The conventional methods of enhancing the rigidity of a component – ribbing, box sections, etc – are all applicable to plastics. Far beyond these, however, modifications such as sandwich moulding, structural foam moulding, and most importantly gas assisted injection moulding, all offer increased stiffness.

Reinforcement in the form of continuous fibres or glass mat is increasingly used as a means of locating rigidity precisely where it is needed: there may be an attendant risk of distortion, needing to be compensated.

Avoiding Brittle Failure

The consequences of unexpected brittle failure in a component can be catastrophic. It happens because unreasonable demands have been made on the material: the usual cause is a lack of consideration for the long chain molecules.

The toughness, or impact strength, of a material is a measure of its ability to absorb received energy and convert it into internal energy (associated with molecular movement). Brittleness can be induced by low temperature or by a high strain rate. At low temperature there is little chance of molecular movement; at high strain rate there is not enough time for movement. At both extremes the polymer is unable to absorb the impact, so that it cracks or shatters. All materials, whether normally tough or brittle, undergo a tough-to-brittle transition as the temperature is lowered. Increasing the rate of strain induces a similar transition. Unfortunately these transitions are rarely very precise, so that there is a zone of uncertainty between the two extremes.

These factors are largely out of the designer's control. All he can do is to ensure that the material is not being asked to perform under unreasonable temperature and impact conditions, and avoid unwarranted stress concentrations. These are the factors which inhibit or prevent molecular movement, over and above the unavoidable constraints of temperature and strain rate. Not that the designer is always to blame. Some of the 'stress raisers' occur during the polymer compounding; others are certainly provided by the processor. Table 8.1 shows some well known examples and some of the correctives.

Table 8.1 Stress concentrations: the causes of disaster

Stress raiser	Corrective
Poorly dispersed pigments or fillers	Extrusion compound
Oxidised polymer particles	Melt screening during extrusion
Voids	Higher injection pressure
Unradiused internal angles	Radius at design stage
Sudden changes of thickness	as above
Uncontrolled weld lines	Mould design; higher temp.
Surface cracks	Various; eg slower cooling

One of the most common design errors is the assumption that form determines function: the thought that the dimensions and configuration which have worked for a familiar material will necessarily work for an unfamiliar one. In the past this has given us brittle door hinges and disposable cutlery which fractures in mid-picnic. Hopefully the plastics industry has put these days behind it.

GOOD PROCESSING

Good processing is essential if plastics are to perform to best advantage. This very inadequate title 'Plastics' is at least accurate when it describes the forming processes. In every process, some combination of heat and pressure is used, first to make the material plastic and then to shape it into its final form. Often the plasticisation and forming must take place within quite narrow limits of temperature, pressure and timing.

Bad processing can exacerbate the effects of bad design: awkward configurations such as very long, thin sections, or severe changes of thickness can constrain the processing parameters very drastically. However even when the design is beyond reproach, bad processing can produce disastrous results. The most familiar faults are the ones associated with the big volume, high speed processes, which involve the rapid heating of material and its shaping under high pressure.

Injection moulding is the most versatile and the most important of the high-speed processes, and arguably the most easily abused. Many problems

Fig. 8.1 Flash in mouldings

can be attributed to 'corner-cutting' in the interests of cost saving. Too much heat input (to assist mould filling) can degrade the polymer: insufficient heat or pressure can produce incomplete 'shorts' or 'sink marks', or, worst of all, voids. These unseen defects may be potentially as catastrophic in critical plastic mouldings as the 'blow holes' were (famously) in the iron castings of the Tay Bridge, which collapsed under wind stresses in 1879. Excessive pressure on the other hand can cause 'flash' (very familiar to users of polystyrene modelling kits), although the prime reason is more likely to be poorly-finished or worn tooling.

Extrusion processes depend on consistency of conditions and uniformity of the melt. Faults in extrusion processes for products like packaging film may be the result of strain in the melt, following damage to the die or uneven extrusion conditions. The defect will be longitudinal, and the consequence could be the embarrassing shedding of a bag-load of wine bottles.

Computerised process control, often with 'closed loop' systems, has revolutionised these high speed operations in recent years. The glaring defects due to poor processing or bad tooling are very rarely seen today, outside the Christmas cracker circuit. Quality control systems are in place to suppress any tendency to over-zealous corner cutting.

COEXISTING WITH METALS

The best solution to many design engineering problems is to use metals and plastics together. As always, coexistence brings its problems, and there are some which cannot be ignored.

Differential Thermal Expansion

In general plastics expand with temperature much more than metals, (although certain thermosetting polymers are an exception). The general case which causes the most trouble is when a plastic shaping is required to perform as the airtight cover or 'lid' on a metal one. A popular solution in packaging is to use a flexible polymer as the lid on a more rigid base and achieve a seal with an interference fit. However when the 'lid' is required to be a rigid composite, a snug interference fit is not an option.

Early attempts in the American automotive industry to use glass reinforced nylon for the rocker covers on metal engine blocks were a typical example. The composite expanded more than the metal (increasingly so at higher temperatures); furthermore its expansion was greater around the bolt-holes, where the fibre orientation was less than elsewhere. The result was oil leaks between the fixing points. The problem has been solved in some cases by using thermosetting materials with lower expansion, but more by redesigning the gaskets to cope with differential expansion. The success of the cooling radiator end tank, also in glass reinforced nylon, over the last 25 years, owes much to the generously proportioned synthetic rubber gasket which marries the end tank to the aluminium base plate.

Anisotropy

The above examples remind us that in reinforced composites, anisotropy affects not only the original dimensions through varying shrinkage, but also the working dimensions through varying thermal expansion. Ideally, the best solution is to design out the distortion tendency. This can be done either by making the shape as symmetrical as possible to minimise orientation, or (if the shape allows), by deliberately encouraging orientation by gating at one end. The aim should be to avoid a situation where the orientation can vary with the processing conditions.

Contact Between Moving Parts

In light weight systems of gears and bearings it is often possible to achieve a better assembly with a combination of metal and plastic parts than either material alone. In assemblies like windscreen wipers, speedometers and film projectors, the benefits of plastics are in reduced maintenance, longer life, and quieter running.

Using Metal Inserts

The pioneers of nylon as an engineering material were quick to appreciate the advantages of incorporating metal inserts in plastic mouldings. Inserts of brass, steel or aluminium are now used in many ways, eg:

- To increase rigidity (especially torsional).
- To increase the pull-out strength of self-tapping screws.
- To provide fixture points for sub-assemblies.
- To permit higher temperature operation in bearings
- To reduce the overall thermal expansion of gear wheels, whilst retaining superior frictional and accoustic performance.

Robots are now widely used to locate metal inserts, in processing SMC and GMT and in blow moulding and rotational moulding, as well as in injection moulding. Processes can now be more accurate and consistently faster.

Coating Metals with Plastics

There is a huge range of applications, from 'high-tech' to 'low-tech', which only exist because of the ability of metals and plastics to work together: one ingredient without the other would be virtually meaningless. Some examples:

- Dip coating with PVC is well established for such familiar objects as refrigerator shelves and sink draining racks.
- Powder coating is used with a variety of polymers to produce thinner but more durable coatings. Fluidised bed coating is a more precise variant, used for such things as steel chair frames and ships fittings.
- Sintering with PTFE is a highly specialised technique which has become very well known through its application to 'non-stick' saucepans and frying pans.
- Wire coating by passing the wire through an extruder head is a very old technique which developed into the cable manufacturing industry. Some very sophisticated products were achieved, like the transatlantic telephone link. The cable industry still prospers, but many telecommunications applications have been taken over by fibre optics.
- The humble twist tie for bag closure is a splendid example. Usually comprising multiheaded coating of thin steel wire with polypropylene, this is a highly successful application which would have been impossible with either ingredient on its own.

JOINING WITH POLYMERS

In the real world of working components, the success of a material frequently depends on how well it can be joined to others. The question is not only are the joints strong, but can they can be made by high speed assembly? There are three basic methods of assembly: by adhesives, by welding, and by mechanical fastening.

Adhesives

Virtually all adhesives are themselves polymers. This is true for the old biologically based gums and resins as it is for the latest wholly synthetic creations. However they are not classified as plastics.

Adhesives are most likely to work with polymers when there is some chemical compatibility with both surfaces, and when there is not too much disparity in rigidity between the two materials. Detailed information is now available from computerised databases, for the three main categories of adhesives for plastics:

- Solvent adhesives: These are most likely to be effective with chemically vulnerable polymers, primarily the amorphous thermoplastics. Most involve rubber, which forms a gap-filling joint as the solvent evaporates. Polystyrene cement for polystyrene modelling kits is a familiar example. Polyvinyl acetate emulsions are similar in that they function by evaporation of water: they are well known on the domestic scene, being satisfactory for making flexible bonds between porous surfaces such as wood, paper and fabrics.
- Hot melts are based on polymers such as LDPE and PET, which are applied to the surfaces with pressure during the cooling cycle. Their main use is in rapid assembly of lightly loaded structures.
- Reaction cured adhesives are thermosetting polymers (usually in the form of two low viscosity ingredients). The modern powerful adhesives are all of this type. Phenolic and epoxy based adhesives are the oldest established, giving very strong bonds but with low flexibility. Two more recent additions, both based on acrylic polymers, are anaerobics, used mainly in thread-locking and pipe-sealing applications; and cyanoacrylates, which set very rapidly into immensely strong bonds, and are most used in high speed assembly of electrical and electronic components. A new dimension of impact strength has been provided by assembling with toughened acrylics and epoxies, incorporating low molecular weight rubbers. These materials are already in use for critical structural applications in railway carriages, vehicles and aircraft.

Welding

Thermoplastics dominate in the world of non-metallic welding. They can be melted and remelted, and precise location of the melt now allows easy assembly of parts.

Welding technology is an area of very rapid change and specialised equipment, but the principles are simple enough. The necessary heat is generated internally or externally by a variety of methods. The choice depends upon the properties of the polymer and the shape and form of the article. The best known processes are these:

- Friction welding takes the form of spin welding for concentric parts, such as collars on pipes. Vibration welding is used for flat surfaces on large mouldings, notably automotive fuel tanks and intake manifolds and radiator expansion tanks.
- Ultrasonic welding is the favourite technique for the mass production of small components like film cassettes and ball point pens, and larger and more intricate shapes like vacuum cleaner bodies and instrument panels.
- High frequency welding is effective for certain polymers; it is very widely used for assembling articles in plasticised PVC, such as inflatables, book covers, wallets and rainwear.
- Induction welding involves generating heat by passing current through a wire or strip pressed between the two parts. It has been used for very large assemblies like the hulls of small boats.
- Hot tool welding uses a plate which is heated and then withdrawn to allow the two parts to be pressed together. This can be a slow process for massive sections such as gas and water mains in polyethylene, but much more rapid for thinner sections. In the automotive industry it is used to join two different materials – ABS and PMMA – for rear light assemblies, and to fabricate radiator expansion tanks from a pair of polypropylene mouldings.
- Hot gas welding uses a welding gun to heat the two surfaces, together with a filler rod in the same material. In large area operations like the lining of reservoirs with polyethylene film, hot gas is applied to overlapped film, without welding rod.

Fig. 8.2 Expansion tank assembled by welding two mouldings in polypropylene

Mechanical Fastening

All manner of mechanical fixing methods have been used for plastics assemblies. To some extent, the rapid progress in adhesives and welding has been made at the expense of mechanical fixing, which tends to be more labour intensive, and is always liable to produce localised stresses. The DIY stores contain a prolific collection of plastic fastening systems.

Fig. 8.3 Wall plug in nylon 6

Self-tapping screws are used with semi-crystalline polymers like polypropylene, acetal and nylons. Low cost sprags and staples are frequently used in automotive assemblies such as polypropylene heater boxes, which are not required to be hermetically sealed. Easy assembly was a big factor in the success of the glass reinforced nylon radiator tank, Fig. 8.4. The traditional braising and welding of the copper–brass radiator was replaced by direct crimping of the aluminium base plate over the flange of the nylon tank, (with the aid of a generous gasket!)

Snap-fitting assemblies are increasingly popular, especially when a single polymer is involved. Unfilled nylon has just the right sort of balance between creep and toughness to make excellent snap fitting assemblies. Electrical connectors for circuitry in the automotive and appliance industries make extensive use of this, Fig. 8.5.

All change!
Having devoted many years to making plastic assemblies secure and permanent, designers are now having to ensure that they can be easily dismantled for segregation, recovery and recycling. This could mean a complete rethink of all plastic assembly operations. It may be that in some cases the twin objectives of secure assemblies and recyclable assemblies will prove to be incompatible.

Fig. 8.4 Radiator tank, showing assembly by crimping

Fig. 8.5 Snap-fitting connectors in unfilled nylon

FINISHING OPERATIONS

Why Paint Plastics?

In the days when they were first being considered for automotive body panels, it was often said that plastics foretold the end of the automotive paint shop. Plastics after all were free from corrosion, could be coloured all the way through, and with the right tooling could be made with high gloss surfaces.

Not so! It soon became apparent that the weathering performance of plastics left much to be desired. Most materials showed fading, loss of gloss, staining and dirt retention. Whilst PVC rainwater goods and window frames and acrylic road signs can keep their good looks for many years, no untreated plastic surface can compete with a properly stoved high gloss car body paint.

Since the early 1980s, the motor industry has been the driving force behind the painting of plastics. The market demands 'Class A' finish for the car body and, increasingly, for the exterior trim. The economic arguments in favour of using more plastics in cars were so strong that satisfactory painting systems compatible with the assembly line became a major research target.

Painting: the Problems

These can be grouped under four headings: adhesion; temperature; performance, and appearance.

Adhesion

Each polymer-paint combination is unique: what works for one system will not necessarily work for another. A slight solvent action can be beneficial, but a clean, degreased surface is always essential. Some polymers, particularly the polyolefines, have an unreactive 'non-polar' surface, and until the early 1980s were regarded as unpaintable in practical terms. Systems have been developed to improve adhesion by rendering the surface more reactive, such as flame treatment, cold gas plasma treatment, and the use of adhesion-promoting primers.

Fig. 8.6 Rover 200 series, 1984: the first example of a painted polypropylene bumper, made possible by an adhesion-promoting primer (courtesy ICI)

Temperature

For the automotive industry this is the dominant issue. The highest temperature experienced on the assembly line is the primer-surfacer 'electrocoat' stage, which can be as severe as 220 °C for 30 minutes. None of the traditional plastics can meet this requirement, so it has been necessary to add any plastic panels further down the line. However the development of a new polymer blend by GE Plastics has overcome this problem. In this the continuous phase is nylon 66, a semi-crystalline polymer with a melting point over 260 °C. The necessary rigidity is provided by the disperse phase of PPO, an amorphous polymer with a suitably high transition temperature. This material was introduced in the GM Saturn concept car in 1990, for the front fenders and rear quarter panels (see Fig. 4.7). By 1997 versions of the new blend had appeared in several production models, notably from Renault. The refinement of incorporating conducting ingredients to make the material suitable for electrocoat has since been added. This meets the once 'impossible' target of using a single painting system for the whole hybrid steel and plastic body.

Performance

One of the great advantages of plastics is their damage tolerance. In automotive terms this translates into the ability of bumpers and wings to sustain small impacts without visible change. For a painted panel there are two crucial requirements: both paint and substrate shall be undamaged, and the adhesion must not be impaired. If the paint layer is significantly more rigid than the substrate, and the adhesion is good, a crack in the paint can initiate a crack in the substrate. There were cases in the 1980s where enthusiasm for pushing plastic bumpers down the complete paint line resulted in brittle bumpers. The solution, now widely practised, is to use a flexible primer coat between the resilient plastic bumper and the rigid topcoat.

Appearance

The motoring public makes no concessions: the painted plastic panels must be indistinguishable from the steel ones. Hence there have been problems with polymers which release traces of volatiles during baking, and with glass reinforced substrates which reveal an uneven surface on cooling. However this remains something of a fashion feature, however universal it may be. A matt automotive finish may come to be desirable, as it has already in many types of appliance.

OTHER SURFACE TREATMENTS

In-mould coating
This tends to be used for large parts not required in very large numbers: it is particularly effective with polyurethanes.

In-mould decoration
This technique has been used for many years in compression moulding, by inserting a decorated foil in the tool.

Printing
Materials like polyolefines are not affected by printing ink solvents, so surface treatments such as flaming or photosensitising are necessary before printing.

Transparent coating
Coating of acrylic and polycarbonate glazing improves the abrasion resistance of the surface and enables these polymers to compete with glass. An important recent development is the coating of polycarbonate headlamp lenses with silicones.

Plating
Three stages are needed in order to produce a chrome finish on plastics. The first is chemical etching, the second is chemical deposition of metallic copper to provide a conducting surface and so permit the third stage, of conventional electroplating. ABS is generally the most successful base for chromium plating.

Vacuum metallising
This is the preferred technique for applying a metal finish, although the coatings are much thinner and less durable. Excellent optical effects are obtained, particularly with acrylics, by second surface metallising, where the coating is on the reverse (unexposed) surface.

Electrostatic flocking
This is a technique for imparting a velour-type finish to plastic film, for decorative or acoustic effects.

Laser marking
A new technique for the engraving of plastics, this must have very considerable potential. It is effective for all thermoplastics, needs no special cleaning or preparation, and (once the up-front costs have been absorbed) it is a cheap, low maintenance operation.

These finishing techniques are not simply cosmetic 'add ons', which will work with any polymer surface. Their importance now is such that how effectively they work can determine the choice of material.

9. FUTURE CONDITIONAL
Challenges and Opportunities

SUCCESS STORY

The growth of the plastics industry has been phenomenal. From its various roots in the nineteenth century, it only began to become commercially significant in the 1920s. Technical innovations in the Second World War and the development of cheap oil ushered in a period of explosive (and at first, chaotic) growth in the 1950s and 1960s. Annual world usage now exceeds 130 million tonnes, and there is little sign of a levelling-out.

Two basic truths have been stated in this book, and will bear repeating. The first is that polymers (like other materials) are not the initiators of progress. They are the servants of the user industries, and they are useful only when they can be processed efficiently and economically. The second is that all our current activities have their roots in Man's compulsion to master his environment. The post-war growth curves may be steeply exponential, but they are smooth curves, developing from recognisable beginnings. The surge in activity is partly the result of the availability of cheap and reliable feedstock from oil and gas, but also the product of colossal world-wide input of research and technology at every stage in the supply chain.

It is worth recalling that the plastics industry's dependence on oil is quite recent. DuPont's first nylon production was based on coal tar, and ICI's first polyethylene plant was fed on ethylene derived from alcohol by fermentation. The limitations of coal as a source of pure polymer feedstock were already becoming apparent in 1939. Once the basic technology of oil and gas refining was established, there was really no contest. Oil and gas, compared with coal, yield monomers more cheaply and more easily, offer more flexible processes, and present fewer environmental problems.

The plastics industry is essentially dynamic. Even apparently 'mature' sectors are always subject to change. Wherever there is competition, the search for cheaper or more effective alternative plastics is relentless. So great is their versatility that there is nearly always a new solution to be found, either in design, or process, or material.

TOO MUCH CHANGE, TOO QUICKLY?

The affluent among us live in a period of unprecedented prosperity. Our living standards would have needed whole armies of servants to maintain, only four or five generations ago. Not surprisingly, many of us are uncomfortable about the modern world and its rate of change. We have strayed too far from our 'natural' beginnings, (so the thought goes), and we shall suffer for it. These guilt feelings co-exist with an appetite for 'shock horror' stories, which our well-developed media are ever eager to satisfy. Unfortunately the technical excellence of the media coupled with the anti-science, anti-industry feelings of the targeted public do not encourage a balanced all-round appraisal. Plastics are a particular casualty of this imbalance, because, compared with more familiar materials, they come wrapped in ignorance.

New dangers to health and threats to the environment are signalled continually. Time and again the villain is identified as the chemical industry in general, and often the polymer industry in particular. Concepts like food additives, detergents, new packaging and new materials are all regarded as 'synthetic', and therefore unnatural. Certainly, some innovations could have far-reaching and possibly hazardous consequences: others are simply novel. The problem with our society is that we are insufficiently educated to distinguish between the two.

In fact all these 'new' phenomena have their roots in old established practices. The differences now are in the scale and particularly the rate of transformation: huge changes can happen within the span of one working life. There are now measurable differences in what we thought was a fixed unalterable background. Notably. the realisation that we can no longer take our climate and our water supply for granted is disturbing indeed.

This situation encourages us to opt either for tradition or innovation (although today the conflict is usually over-simplified into Green or anti-Green). Heinz Gartmann, writing in the 1950s, spoke of 'the eternal conflict between the innovator and the traditionalist'. Of course without the innovators the world would have stood still; but he also emphasised that without the opposition of the conservatives there would be chaos. In short, we need balance. We within the industry should recognise that opposition is not only natural, but necessary. Imagine a world without protective legislation for food, buildings, sewage, transport and the like, and a supply chain for the automotive and appliance industries unrestrained by specifications or quality control. These checks and balances are essential if technological disasters are to be avoided. The environmental chaos left by the command economies of Eastern Europe is there to prove it.

Opposition to chemicals and plastics can be focussed into three areas:

Atmospheric pollution and climatic change
Threats to health
Poor management of resources and waste.

(In this context it is not always possible to draw a clean line between 'chemicals' and 'plastics'. A great variety of chemicals is involved in the production and also in the decomposition of plastics; and plastics, after all, are chemicals.)

ATMOSPHERIC POLLUTION AND CLIMATIC CHANGE

The Unclean City

In the 'Good Old Days' atmospheric pollution was all too visible. For many centuries charcoal burning was a considerable nuisance in rural areas. Much worse later was coal burning, as a source of power in factory and power station and as the main source of domestic heating. Continual severe fogs were accepted as part of the urban scene for over a century. The evil effect of London 'pea-soupers' on the health of its citizens was accepted, along with the despoiling of buildings and washing lines. Clean Air legislation and the introduction of smokeless fuel in the 1950s effected a big improvement surprisingly quickly.

Atmospheric pollution in the city is nowadays centred on exhaust emission from motor vehicles. The chief undesirables are carbon monoxide, nitrous oxide, and unburnt hydrocarbons. Plastics help in two ways towards reducing the problem. Catalytic converters (with housings usually moulded from glass reinforced nylon) remove between 75% and 90% of these gases. Furthermore, the overall weight reduction of vehicles made possible by plastics is already reducing the total exhaust gas emission by at least 5%.

Factors like these however are only tinkering with the problem. Clean air in the cities can only be achieved by eliminating fossil fuel burning vehicles from them. Slow moving, stop-start traffic is in any case a grossly inefficient and polluting consumer of fuel.

The search for alternative fuels is now being pursued (rather belatedly, and rather more in the USA and Japan than in Europe). The motor manufacturers are investing millions in the development of electric vehicles and (more optimistically) hybrids, which use batteries in the city and petrol out-of-town. These new classes of small vehicles could establish an entirely new culture of performance and finish; practical and realistic, less dominated by saleroom fantasy and the company car. Whichever way the popular taste develops, there is no doubt that it will prove an immense boost for new body designs in plastics.

Whether we incline towards the Brave New World or the Good Old Days, our preoccupation with exhaust emissions from motor vehicles should not obscure the fact that city streets in the days before the car were extremely unpleasant places. In London in the 1880s there were over 100 000 working horses: the streets of this much smaller London were subjected to over one thousand tonnes of horse droppings each day. This provided work for an army of crossing sweepers. . . . and the smell can only be imagined. Smell was also the dominant feature of the urban rubbish dump, and of the domestic dustbins which fed it. We only began to escape from these horrors with the advent of polyethylene bags and sacks in the 1960s.

Acid Rain

Acid rain was one of the first effects to remind us that pollution does not acknowledge national frontiers. It arises mainly from oxides of sulphur and nitrogen from fossil fuel burning, All of Northern Europe is experiencing dead trees and poisoned lakes. Amongst European acid rain exporters, Britain, invidiously, is high in the league. Plastics however are not significant contributors.

The answer to acid rain is not simply to build higher power station chimneys: this may clean up our own act but causes more acid rain to be exported elsewhere (from Britain this usually means Scandinavia, because of the prevailing wind). The solution is to install flue gas scrubbers in the chimneys. With the right conditions of time and temperature, nearly all the potentially corrosive gases are destroyed. Plastics help this process considerably: nowadays the scrubbers themselves are made of glass reinforced polyester resin, which is much more corrosion resistant than any feasible metal alternative.

Dioxins

A very great deal of attention has been devoted to the role of dioxins in air pollution. The dioxins are extremely toxic chemicals, which are formed when organic materials containing chlorine are burnt. Such materials include wood, paper and garden refuse, but they also include waste plastics. This has been used as the basis of attacks on PVC, even to the extent of campaigns to eliminate chlorine from all industrial processes. However, exhaustive tests involving burning municipal solid waste in many countries have established that properly run incinerators and scrubbers reduce the dioxin content to well within acceptable limits. Key experiments were conducted in 1995 at the South East London Combined Heat and Power 'Energy from Waste' plant at Deptford. It was demonstrated that adding plastics (including PVC) to the the normal feedstock of municipal solid waste produced no increase in the level of dioxins and other undesirables. These were still within the authorised limit by a factor of 50 or more.

Ozone Depletion

Ozone depletion is now a major environmental issue. The ozone layer in the upper atmosphere is vital to life on earth, because it screens out most of the sun's harmful ultraviolet radiation. Since the mid-1980s it has become clear that the ozone layer is showing seasonal thinning over the polar regions. Chlorine was again identified as the villain of the piece, because of the ease with which it destroys ozone, causing it to revert to oxygen. The main source of chlorine in the upper atmosphere appears to be the chlorofluorocarbons, CFCs. These have been used since the 1930s as refrigerants, and more recently as aerosol propellants.

The relevance of CFCs to plastics is that they are formed in the manufacture of flexible polyurethane foams, and also provide most of the insulating effect in the rigid foams of refrigerator linings. International agreement to phase out the use of CFCs generated exhaustive research programs throughout the 1990s. These have yielded effective (if expensive) replacements for CFCs. In the UK, CFCs were virtually eliminated from aerosols between 1987 and 1990, and from polyurethane foam by 1994. However existing CFCs are likely to be affecting the ozone layer for at least another century. Preventing the escape of CFCs from badly maintained and end-of-life refrigerators around the world will be a huge and continuing challenge.

Global Warming

Of all the observed environmental trends, global warming is potentally the most serious. The earth is maintained at stable temperature by the 'greenhouse gases' in the upper atmosphere. Radiation from the sun is reflected back from the earth's surface: how much is retained and how much escapes back into space is determined by the greenhouse gases. Too much of them means that too much heat is retained, and global warming ensues. The total picture is more complicated, largely because of the balance between polar icecaps, sea levels and water vapour in the atmosphere, and the effect of changes in this balance on the ocean currents. At all events, quite small changes in the global temperature could seriously destabilise climatic conditions. The effect is real enough: the causes are not proven, but several processes stand accused.

The principal man-made additions to the greenhouse gases are CFC's (but not chlorine itself); methane (from ruminating cattle, themselves increasing because of intensive farming); and carbon dioxide. This last is by far the most abundant. Every year around 20 billion tonnes of carbon dioxide are generated by fossil fuel burning. Half of this comes from power stations, about 10% from burning rain forests, and much of the rest from motor vehicles.

Plastics could play a vital role in reducing carbon dioxide evolution, mainly through lighter vehicles and better insulation of buildings; but as yet the process has barely begun.

THREATS TO HEALTH

Quality and Quantity

There will always be apprehension about new materials and new processes. As we have noted, health concerns about chemicals and polymers are sensible and desirable. After all, these substances are sometimes made with the aid of intermediates which are unpleasant, and many of them are composed of elements which in other combinations can be lethal. So here are good grounds for vigorous and informed debate.

Unfortunately, there is little debate: there is only abuse and misinformation. The greatest ignorance concerns the difference between quality and quantity. Some substances are indeed harmful even in very small traces: others only in very large quantities. Taken in sufficient quantity, anything can be harmful. This concept has been appreciated by scientists for centuries. In the sixteenth century, the physician and alchemist Paracelsus wrote;

> *'All things are poisons; nothing is without poison; only the dose determines whether there is a harmful effect.'*

Alas, this simple idea has not taken root in the community at large. Its absence is enthusiastically exploited by the 'Anti' lobby. Effects from massive doses are extrapolated down to minute ones, and from one species to another very different one. The argument says, in effect, that there is no such thing as a safe dose. Our news media, once again, are very skilled and very active in picking up and transmitting signals of disaster. Denials and reappraisals are less actively pursued; they have negligible news value.

The balanced view incorporates the concept of Risk Management. Here risk is defined as the product of hazard and exposure. A slight hazard could result in significant risk of harm if there is widespread exposure to it. However, even if a hazard is very severe, if people are fully protected from it, ie the exposure is zero, then there is no risk.

There follows a summary of the the main health issues which face the polymer industry today.

Carcinogens

Many chemicals in common use today have been at some time accused of being carcinogens. The allegations are often based on the observed effects of massive doses in laboratory rats. Charges of this kind have been made against DEHP phthalate plasticisers in flexible PVC, and against phosphate fire retardant additives in polyurethane. Neither has been substantiated. The one significant problem that has been identified concerns vinyl chloride monomer (VCM). In the 1970s a rare form of liver cancer was identified in people working in PVC polymerisation plants, who were exposed to heavy concen-

When the weight of an assembly is relevant, the balance shifts more drastically in favour of plastics. Plastic moving parts use less energy.

Dow Plastics have quoted the example of a car fender (wing) made respectively in steel and polyurethane. The plastic fender is 2.8 kg lighter, which translates into a saving (over 150 000 km) of 25.2 kg of fuel. This energy saving in use is around 3.6 times the energy expended in production. Aluminium, although offering a weight saving of 2.4 kg, can never save enough energy in use to compensate for the very high energy used in extraction.

Plastics Waste

'Plastics packaging is the prime example of the throw-away society: it isn't reusable, it isn't recyclable, and it makes up most of our domestic rubbish'.

Comments like this are made 'authoritatively' almost every day: they are always wrong. All plastics are technically recyclable (although not always economically), and of the domestic waste stream they constitute at the most 10–11% by weight.

Figures for the composition of the contents of the domestic dustbin in the UK in 1993 are quoted in Table 9.1:

Table 9.1 Contents of the Domestic Dustbin (by weight), courtesy BPF

Paper and card	32.4%
Putrescibles	20.2%
Glass	9.3%
Miscellaneous combustibles	8.1%
Fines	6.8%
Dense plastic	5.9%
Ferrous metal	5.7%
Plastic film	5.3%
Textiles	2.1%
Miscellaneous non-combustibles	1.8%
Non-ferrous metal	1.6%

Recycling

It is curious how much interest there is in preserving the 4% of the oil barrel represented by plastic materials, and how little concern there is about the 80/85% which we burn. The reason is visibility; of packaging, of scrap cars and appliances. We have all grown up in an almost totally indisciplined litter-creating culture. It gets worse because as we become more affluent, the more we throw away. Plastics of course weigh so little and mostly are in the form of film. A little goes a long way.

Traditionally, industry made little attempt to recover waste, unless it was uncontaminated, easily reusable and therefore economically expedient. Now, the pendulum has swung; every company addresses the subject of waste. Several hundred companies have been set up in Europe in recent years, to take in plastic waste and reconstitute it into something useful. They each have a different focus. It is a very active scene; one which is changing very rapidly.

Looking at the whole subject of waste management (not just recycling, which is not always the best option), there are several possible courses:

1. Component re-use: this may be the most economical, unless the component design and specification have changed during the life cycle. (Usually, of course, they have!) The need for cleaning or touching-up could also render this option less attractive.
2. Recycling of material, undiluted and uncontaminated. This is being done quite satisfactorily, for Renault and VW bumpers.

   The retail trade recycles items like polystyrene coathangers and polyethylene film cassettes very efficiently. This is also true of polypropylene battery cases, although Fig. 9.3 shows that these can be very dirty. (Indeed the cost and energy involved in any emptying and washing operations must always be addressed.)

   Clean recovered polymer is already widely remoulded into usable components. The normal pattern is to use the recycled polymer in slightly less demanding applications: there is always the possibility of the material having been somewhat degraded or contaminated. The automotive parts in Fig. 9.4 are a successful recycling example.

Fig. 9.2 Collection of VW bumpers for recycling (courtesy Bird Group)

Fig. 9.3 Polypropylene battery boxes for recycling (courtesy BPF)

Fig. 9.4 Automotive parts moulded from recycled polypropylene

Many supermarket chains now collect the polyethylene pallet overwrap entering their warehouses, for recycling into carrier bags. Every plastics converter (in sheer economic self-interest) makes the best possible use of the offcuts and single-material processing scrap he generates. (But not all such scrap is in a usable physical form; witness the lumps of solidified melt purged from extruders, picturesquely known as 'cowpats').
3. Recycling of mixed plastics. This is something of a jungle, Unspecified mixtures can be difficult (and even dangerous) to process, and their

properties uncertain. The products may be less reliable and more expensive than second-grade polymer. Furthermore, the market for mottled and speckled fence posts, road cones and park benches is not unlimited.

(It will be clear from (2) and (3) that effective recycling from assemblies like motor cars is really limited to large easily-dismantled single-polymer parts. A car has around 1000 plastic components, most of them small parts of sub-assemblies with several materials. Car makers are now under pressure to design for recycling, so that sub-assemblies like instrument panels only involve one polymer family.)

4. Recycling of monomers (in theory a satisfyingly 'pure' solution). There are various methods, notably pyrolysis, of depolymerising polymers and recovering the chemical 'building blocks'. These work reasonably well with a few materials like polyurethane and acrylics, but for most polymers this solution lies somewhere over the horizon.
5. Energy recovery by incineration. One fact is clear. Once plastics waste has become irretrievably mixed or intimately contaminated, it has no future as recycled material. Its value lies in the energy it contains. It has a higher calorific value than coal, and will improve the combustion of the conventional feedstock. Incineration however needs careful management: apart from the question of emissions, operators of power stations and blast furnaces are very particular about what feedstock they use. Specialist combined heat and power plants are needed. As yet not enough of these clean high temperature incinerators are in place, but there can be no doubt that this is where the future lies. On average, incineration of plastics leaves a dense residue with a volume equal to only 3% of the input.
6. Landfill. This option is severely out of favour throughout Europe: it is really only appropriate for concentrated incineration residues. As stable non-biodegradable infill, this is still useful: it is the correct resting place for the 'irreducible minimum' of the recycling sequence. It is much more suitable than our current landfill material, which is unconcentrated, physically unstable, and loaded with methane-generating putrescibles.

CONCLUSIONS

Plastics are deeply involved with all the cutting edges of technology, in information and communication just as in transport and electrical appliances. Their role in the food chain is absolutely vital. Without plastics there would be no way of keeping everyone fed until the population explosion is brought under control (at best, halfway through the century).

The UK government is currently looking at the application of sustainable development strategies for industry. 'Sustainablity' represents the desirable balanced solution, the happy mean between the extremes of unchecked technology and the 'Back to Nature' option. Sustainability, applied to resources, means meeting the needs of the present without compromising the Future. The plastics industry is aligning itself with this concept.

The more one looks at 'the Problems of Plastics', the more they become identified as the problems of Society. Our attitudes towards litter and the squandering of energy are ignorant and irresponsible, but not beyond recovery, given a positive approach to science and technology, and an educated response to gloomy or malicious doom-mongering. People positively enjoy participating in material collection schemes and energy conservation, but they are continually let down by inadequate facilities. The infrastructure must be put in place, and then made to function by whatever sticks and carrots are needed. Then perhaps plastics will be recognised as the prize they really are.

BIBLIOGRAPHY

M. G. Bader: *Polymers for Advanced Applications*, University of Surrey course papers, 1993.
A. W. Birley and M. J. Scott: *Plastics Materials, Properties and Applications*, Leonard Hill, 1982.
D. W. Clegg and A. A. Collyer: *The Structure and Properties of Polymeric Materials*, The Institute of Materials, 1993.
K. Easterling: *Tomorrow's Materials*, The Institute of Materials, 1993.
S. Fenichell: *Plastic: the Making of a Synthetic Century*, HarperCollins, New York, 1996.
H. Gartmann: *Science as History*, Hodder and Stoughton, 1961.
S. Katz: *Classic Plastics*, Thames and Hudson, 1984.
M. Kaufman: *The First Century of Plastics*, The Plastics and Rubber Institute, London, 1963.
J. Martin: *Materials for Engineering*, Institute of Materials, 1996.
J. Maxwell: *Plastics in the Automotive Industry*, Woodhead Publishing, Cambridge, and Society of Automotive Engineers, USA, 1994.
R. Newport: *Plastics Antiques*, British Industrial Plastics Ltd, 1976.
H. J. Saechtling: *International Plastics Handbook*, Carl Hanser Verlag, Munich, 1987.
G. Thomas and M. Stephenson: *Understanding Plastics*, Chemical Industry Education Centre, University of York, and Elf Atochem, 1997.
C. Wilson: *Environmental Questions, and Some of the Answers*, BASF, 1991.
G. Woods: *The ICI Polyurethanes Book*, John Wiley & Sons, 1990.

British Plastics Federation publications:
    Additives make Plastics, 1989
    Can it be Made in Plastics or Rubber?, 1997
    EPS and the Environment, 1992
    Plastics in the Home, 1991
    Plastics on the Move, 1994
    Plastics in Power, 1993
    Plastics, our Versatile Servants, 1987
    The World of Plastics, 1986
    Plastics Waste – A Source of Useful Energy, 1992
    PVC – the Facts, 1994
    South East London Combined Heat and Power Trials, 1995
    Thermoset Plastics Products Recycling, 1993

Index

Abrasion resistance, 101–2
ABS, 27, 55, 122
Absorption, 111
Acetals, 24, 53
Acid rain, 126
Acoustics, 39–40
Acrylonitrile-butadiene-styrene (ABS), 27, 55, 122
Additives, 60–1, 100–1
Adhesives, 116
After-shrinkage, 110
Ageing, 94, 95–8
Agriculture, 7, 35
Aircraft applications, 13, 40, 78, 91, 101
Amorphous polymers, 3, 50, 56, 95
Animals
   hazards of litter to, 130
   products from, 131–2
Anisotropy, 82–3, 110–11, 114
Antioxidants, 96
Aromatic polymers, 53–4, 55, 97, 98, 100, 101
Artificial ski slopes, 26
Assembly methods, 115–18
Automotive applications, 14, 27–8, 33–4, 35–6, 38, 61–3, 89, 92, 114, 125, 133, *see also* Cars
   body panels, 45, 60, 78–9, 80, 85, 86, 91–2
   bumpers, 60, 63, 86, 134
   catalytic converters, 125
   control pedals, 44–5
   drive shafts, 91
   front and rear ends, 14, 28, 35, 36, 88
   fuel tanks, 14, 28, 73, 74
   intake manifolds, 14, 31, 72, 88
   lights, 40, 72, 102, 117, 122
   painted plastics, 120–1
   radiators, 117, 118
   seats, 22, 23
   sound insulation, 39–40
   windows, 40

Baekeland, Leo H., 5

Bag closures, 115
Bag making, 29, 69
Bakelite, 5
Beechcraft Starship, 91
Benzene rings, 53–4
Billiard balls, 5
Biodegradable plastics, 130–1
Blends, 59–60
Blow moulding, 28–9, 73
Boats, 80–1
Bottles, 8, 9, 20–1, 29, 35, 40, 73, 129
Brittleness, 51, 105, 111–12
Bubble-pack, 30
Building and construction, 7, 10–11, 41–2, 92
Bulk moulding compound, 71, 89
Burning plastics, 94, 99–101, 129, 136

Cable covering, 5, 12, 67, 115
Carbon black, 98
Carbon dioxide, 127
Carbon fibre reinforcement, 16, 31, 42, 91, 101
Carbonated drinks bottles, 9, 20–1, 29, 35, 73
Carcinogens, 128
Cars, *see also* Automotive applications
   Chevrolet Corvette, 79–80
   Citroen Xantia, 85
   Formula One racing cars, 91–2
   Lotus, 86
   Renault Espace, 86, 91
   Soya Bean Car, 78–9
   Trabant, 79
Casein, 5
Catalytic converters, 125
Celluloid, 5
Cellulose acetate, 5
Cellulose nitrate, 5
Ceramics, 42–4
CFCs, 127
Chalking, 98
Chemical resistance, 56, 99

Chevrolet Corvette, 79–80
Chlorine, 127
Chlorofluorocarbons, 127
Chrome finishes, 122
Citroen Xantia, 85
Civil engineering, 10–11, 92
Cloth impregnation, 5, 78
Clothing, 5, 7, 38, 78
Coat hangers, 47
Coatings, 115, 119–22
Coloured plastics, 37
    painting, 119–21
Component costs, 33–4
Composites, 61, 77–92, *see also* Fibre reinforcement
Compression moulding, 74
Computer aided design and manufacture, 70, 113
Computer databases, 107
Consolidation, 34–5
Construction industry, 7, 10–11, 41–2, 92
Continuous fibre composites, 84, 90–2
Continuous service temperature, 55, 96
Copolymers, 59–60
Corrosion, 39
Cost factors, 33–6, 55
Creep, 50, 111
Crockery, 43
Crystallinity, 3, 20–1, 49–50, 56, 95
Curtain fittings, 26

Databases, 107
Decorative plastics, 10, 122
Defects in materials, 113
DEHP phthalate plasticisers, 128, 129
Density, 35, 94–5
Dentistry, 15
Design, 36–7, 93–4, 109–12
Design stress, 51
Dioxins, 126
Dispersed fibre composites, 84, 88–90
DIY industry, 11, 47
Domestic waste, 133
Doulton, Henry, 43
Drinks bottles, 8, 9, 20–1, 29, 35, 40, 73, 129
Ductility, 51, 105
Dustbins, 74

'Ebonite', 57
Ecobalances, 132–3

Economic factors, 33–6, 55
Elastomers, 3, 56–7, 75
Electrical applications, 5, 7, 11–13, 38, 43, 67, 89, 100, 103, 115
Electrocoating, 121
Electrostatic flocking, 122
Energy
    consumption, 132–3
    production, 126, 127, 131, 136
Engineering plastics, 53–4, 55
Environmental issues, 17, 125–7, 129–36
Expanded polystyrene, 9, 22, 23, 130
Extrusion, 66–9, 113

Fabric impregnation, 5, 78
Fasteners, 118
Fibre reinforcement, 11, 61, 76, 77–92, 94, 95, 96, 104, *see also* Carbon fibre reinforcement
Filament winding, 90–1
Filaments, 67
Filled polymers, 61
Films, 21, 29–30, 67–9
Fire retardants, 128, 129
Fish processing, 44
Flammability, 94, 99–101, 129, 136
Flash, 113
Flocking, 122
Fluid absorption, 111
Fluoropolymers, 55, 94
Foam moulding, 71
Foams, 9, 21–3, 75, 127
Food packaging, 7–9, 30, 67, *see also* Bottles
Footwear, 23
Ford, Henry, 78–9
'Formica', 10
Formula One racing cars, 91–2
Fracture, 51
Friction, 24–6, 38–9, 101–2
Fuel tanks, 14, 28, 73, 74
Furniture, 7, 16–17
Fusible core moulding, 20, 30–1, 71–2

Gardening applications, 17, 27, 36, 41, 43
Gartmann, Heinz, 124
Gas assisted moulding, 71
Glass fibre reinforcement, *see* Fibre reinforcement
Glass mat thermoplastics, 86
Glass reinforced polyester (GRP), 11, 79–81

Glass replacement, 40–1
Glass transition temperature, 95
Global warming, 127
Goodyear, Charles, 5
Great International Exhibition, 106
Greenhouse gases, 127
GRP, 11, 79–81
Gutta-percha, 5, 19

Halogens, 101
HDPE, 9, 53, 55
HDT, 95
Health hazards, 128–9
Heat, *see* Burning plastics; Temperature stability
Heat distortion temperature, 95
Heat insulation, 38
Heat shields, 97
Heat stabilisers, 96
Heterocyclics, 55, 97
High density polyethylene (HDPE), 9, 53, 55
Hinges, 20
Holoplast, 10
Hormone disruption, 129
Horn, 19
Horticultural applications, 17, 27, 36, 41, 43
Housewares, 7, 16, 26, 43

Impact strength, 105–6, 111
In-mould coating, 121
In-mould decoration, 122
Incineration, *see* Burning plastics
Injection moulding, 20, 30–1, 69–72, 75, 86–7, 112–13
Intake manifolds, 14, 31, 72, 88
Ivory, 5

Just-in-Time, 34

Laminates, 10, 77–8
Landfill, 136
Laser marking, 122
LDPE, 8, 10, 26, 53, 55
Lead replacement, 45–6
'Lego', 27
Lightness, 35, 94–5
Litter, 129–30
Liver cancer, 128
Lotus cars, 86

Low density polyethylene (LDPE), 8, 10, 26, 53, 55
Lubrication, 24–6, 38–9, 101–2

Mackintosh, Charles, 5, 78
Material selection, 106–8
Mechanical fasteners, 118
Medical applications, 7, 15, 99, 102
Metal
    coating with plastics, 115
    combined with plastics, 113–15
    finishes on plastics, 122
    inserts, 114–15
    replacement by plastics, 44–6
Microcrazing, 98
Military applications, 78, 79, 80–1
Milk bottles, 9
Modulus, 50, 94, 95, 96
Molecular structure, 2–4
Molecular weight, 58
Monomers, 3
Mooring buoys, 23
Motor industry, *see* Automotive applications; Cars
Multi-shot moulding, 71

Natural polymers, 4
'Netlon', 29
Non-stick coatings, 53, 115
Nylon, 12, 16, 20, 24, 53, 55, 100
    absorption by, 111
    reinforced, 31, 42, 88–9

Oestrogen mimicking, 129
Oil consumption, 131, 132
Oxidation, 95–6, 97–8
Ozone depletion, 126–7

Packaging, 7–10, 29–30, 67, 114, 129–30, 133, *see also* Bottles
Painting plastics, 119–21
Papier maché, 19
Paracelsus, 128
'Parkesine' exhibit, 106
PBI, 97
PBT, 54, 55
PEEK, 55, 100 101
PES, 55, 99, 100, 103
PET, 9, 20–1, 54, 55, 73, 102
Phenol-formaldehyde resin, 5

Phenolic composites, 77–9, 89
Phthalate plasticisers, 128, 129
Pipes and tubes, 10, 36, 43, 45, 67
Plasticisers, 128, 129
Plastics
   choice of, 106–8
   deficiencies of, 44
   historical development, 5–6
   joining together, 115–18
   molecular structure, 2–4
   opposition to, 124–37
Plating plastics, 122
PMMA, 40, 55, 71, 98, 102
Pollution, 17, 125–7, 129–31
Polybenzimidazole (PBI), 97
Polybutylene terephthalate (PBT), 54, 55
Polycarbonate, 54, 102
Polyether-etherketone (PEEK), 55, 100 101
Polyethersulphone (PES), 55, 99, 100, 103
Polyethylene terephthalate (PET), 9, 20–1, 54, 55, 73, 102
Polyimide, 55
Polymers, 2–4, 49–63, *see also* Plastics
Polymethyl methacrylate (PMMA), 40, 55, 71, 98, 102
Polyolefines, 53, 100, 120, 122
Polypropylene (PP), 16–17, 20, 37, 43, 53, 55, 61–3
Polystyrene, 9, 22, 23, 102, 130
Polytetrafluoroethylene (PTFE), 24, 52–3, 103, 115
Polyurethane foam, 23, 75, 127
Polyvinyl acetate emulsions, 116
Polyvinyl chloride (PVC), 10, 11, 41–2, 43, 53, 94, 98, 115, 117, 126, 128–9
Powder coating, 115
Power plants
   fossil fuelled, 126, 127, 131
   plastic incineration, 136
Printing on plastics, 122
Processing, 65–76, 112–13
Protective clothing, 38
PTFE, 24, 52–3, 103, 115
Pultrusion, 89, 90
PVA adhesive, 116
PVC, 10, 11, 41–2, 43, 53, 94, 98, 115, 117, 126, 128–9

Racing cars, 91–2
Railways, 13–14

Rain erosion, 98
Raincoats, 5, 78
Rainwater goods, 10
Random mat composites, 84–8
Raw material costs, 33
Reaction injection moulding, 75, 86–7
Recording industry, 21
Recycling, 133–6
Refrigerators, 23, 127
Refuse sacks, 29, 69
Reinforced plastics, *see* Fibre reinforcement
Renault Espace, 86, 91
Resin transfer moulding, 86
Resource depletion, 131–6
Rigidity, 50, 93–4, 104–5, 111
Risk management, 128
Rotational moulding, 73–4
Rubber, 5, 19, 57

Sack making, 29, 69
Safety, 38–9, 128–9
SAN, 55, 102
Sandwich moulding, 71
Sanitary ware, 43
Self-tapping screws, 118
Semi-aromatic polymers, 54, 55, 100
Semi-crystalline polymers, 3, 20–1, 49–50, 56, 95
Semi-synthetic polymers, 4
Sewage pipes, 43
Sheet moulding compound, 85–6
Shellac, 5
Ships, 80–1
Shorts, 113
Shrinkage, 110
Sink marks, 113
Snap-fittings, 118
Sound insulation, 39–40
Soya Bean Car, 78–9
Space vehicle heat shields, 97
Sports goods, 7, 15–16, 23, 31, 42
Staudinger, Hermann, 2–3
Steel replacement, 44–5, *see also* Metal
Stiffness, 50, 93–4, 104–5, 111
Strain, 50, 51, 111
Strength, 50–1, 93–4, 104–5
Stress, 50, 51, 99, 112
Styling, *see* Design
Styrene-acrylonitrile (SAN), 55, 102
Surface treatments, 115, 119–22

Sustainability, 136–7
'Swappit', 46
Synthetic polymers, 4

Temperature stability, 25, 55, 95–7, 110, 114
Tennis racquets, 31, 42
Tg, 95
Thermal insulation, 38
Thermal stability, 25, 55, 95–7, 110, 114
Thermoforming, 76
Thermoplastics, 4, 51–6, 57, 95, 100
Thermosets, 3–4, 52, 55, 57, 95, 100
Thickness, 36, 93–4
Timber replacement, 41–2
Toughness, 105–6, 111
Toys, 7, 16, 26–7, 37, 46
Trabant, 79
Traffic cones, 74
Transparency, 54–5, 56, 94, 102
Transparent coatings, 122
Transport applications, 7, 13–14, *see also* Aircraft applications; Automotive applications; Boats
Tubes and pipes, 10, 36, 43, 45, 67
Twist ties, 115

Ultraviolet (UV) ageing, 97–8

Vacuum forming, 76
Vacuum metallising, 122
Vinyl chloride monomer (VCM), 128–9
Visco-elastic behaviour, 50
Voids, 113
'Vulcanite', 57

Wall thickness, 36
'Warerite', 10
Washing machines, 35, 88
Waste management, 133–6
Wear resistance, 101–2
Weight, 35, 94–5
Welding, 116–17
White goods, 35, 88, 127
Window frames, 10, 41–2
Windows, 40
Wire coating, 115
Wood replacement, 41–2
Woven fibre composites, 84, 91–2

'Xylonite', 5

Yielding, 51
Yogurt pots, 30